KUMON MATH WORKBOOKS

Word Problems

Table of Contents

1 Review

Date / / Name

Level ☆ Score /100

1 4 boys and 5 girls are sitting in class. How many children are there altogether? 10 points

⟨**Ans.**⟩ ______

2 I have 7 pencils. If Tim gives me 5 pencils, how many pencils will I have in all? 10 points

⟨**Ans.**⟩ ______

3 Your mother left 10 apples in the kitchen. You ate 3 apples today. How many apples are left? 10 points

⟨**Ans.**⟩ ______

4 Sue has 6 red flowers and 11 white flowers in her garden. How many more white flowers than red flowers does she have? 10 points

⟨**Ans.**⟩ ______

5 The teacher has 8 apples and 4 oranges. Does he have more apples or oranges? How much more? 10 points

⟨**Ans.**⟩ He has ☐ more ☐.

6 Bob has 6 pencils and his sister has 8 more pencils than him. How many pencils does she have?

10 points

〈**Ans.**〉

7 9 boys and 6 girls are sitting in class. How many chairs do we need if each person wants to sit down?

10 points

〈**Ans.**〉

8 Your mother baked 15 cookies. You ate 8 cookies today. How many cookies remain?

10 points

〈**Ans.**〉

9 A bus carrying 6 people goes to two bus stops. At the first stop, 3 people get on. At the second stop, 5 more people get on. How many people are on the bus altogether?

10 points

〈**Ans.**〉

10 I have 10 books on my bookshelf. To the right of my book about cars there are 5 books. What number book from the left is my book about cars?

10 points

〈**Ans.**〉

Do you remember these problems? Good!

2 Review

Date / / Name

Level ☆ Score /100

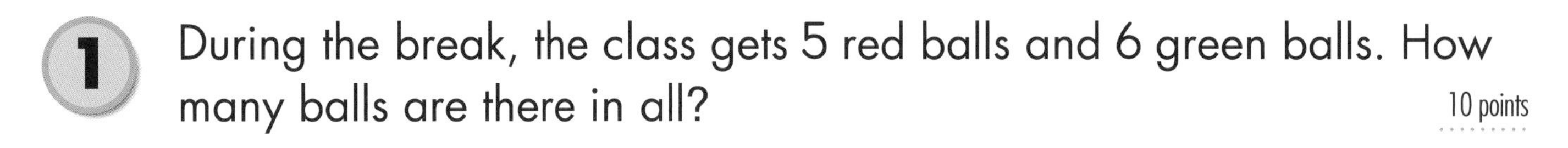

1 During the break, the class gets 5 red balls and 6 green balls. How many balls are there in all? 10 points

〈**Ans.**〉

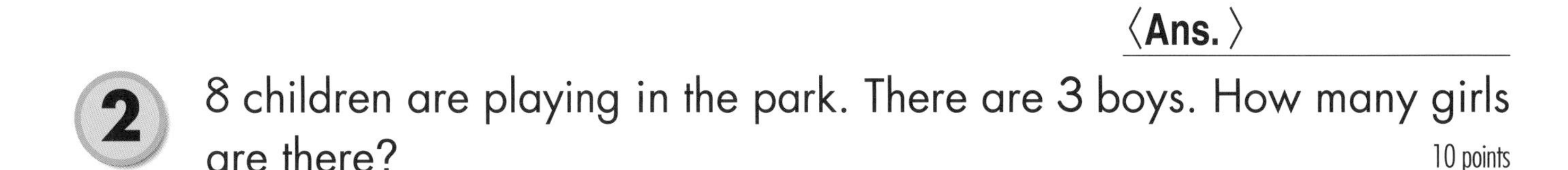

2 8 children are playing in the park. There are 3 boys. How many girls are there? 10 points

〈**Ans.**〉

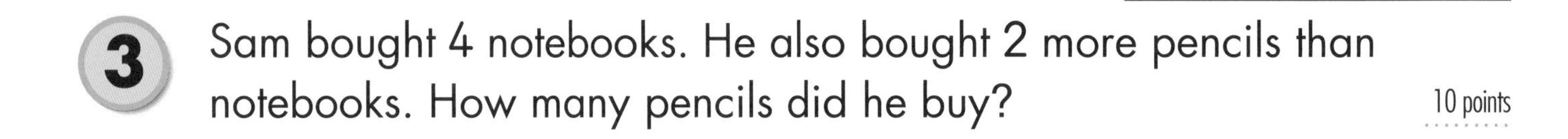

3 Sam bought 4 notebooks. He also bought 2 more pencils than notebooks. How many pencils did he buy? 10 points

〈**Ans.**〉

4 Mark is waiting in line for the bus. There are 3 people in front of him. What number person from the front is Mark? 10 points

〈**Ans.**〉

5 In the fridge, there are 8 bottles. 3 of them are empty. How many full bottles are there? 10 points

〈**Ans.**〉

6 In class, we have 9 chairs. If 4 children each sit down on a chair, how many chairs remain? 10 points

⟨**Ans.**⟩

7 The baker made 4 cakes. He also made 5 more cookies than cakes. How many cookies did he make? 10 points

⟨**Ans.**⟩

8 Caroline has 8 red stickers and 5 blue stickers. How many more red stickers than blue stickers does she have? 10 points

⟨**Ans.**⟩

9 May had 4 dimes. Then her sister gave her 3 dimes. May gave her brother 2 dimes. How many dimes does she have now? 10 points

⟨**Ans.**⟩

10 In the hall, there are 7 hats on hooks. There are 3 hats to the right of the black hat. How many hats are to the left of the black hat? 10 points

⟨**Ans.**⟩

Ready to start something new? Awesome!

3 Addition & Subtraction

Level ☆☆

Date / / Name

Score /100

1 Lisa read 20 pages of her book yesterday. Today she read 30 more pages. How many pages did she read altogether? 10 points

〈**Ans.**〉

2 Bob has 50 pennies and David has 30 pennies. How many more pennies than David does Bob have? 10 points

〈**Ans.**〉

3 Peter's class has 18 boys. There are 3 more girls than boys in his class. How many girls are there in Peter's class? 10 points

〈**Ans.**〉

4 There are 21 boys in Dana's class. There are 3 fewer girls than boys in her class. How many girls are there in Dana's class? 10 points

〈**Ans.**〉

5 Adam picked 36 strawberries. Later he picked 8 more. How many strawberries did he pick in all? 10 points

〈**Ans.**〉

6 Cathy sent 6 postcards out of the 25 postcards she had. How many postcards does she have left? 10 points

〈**Ans.**〉

7 The pear tree had 45 pears on it last week. 8 new pears grew this week. How many pears are on the pear tree now? 10 points

〈**Ans.**〉

8 The store had 43 apples. A customer left with 7 apples. How many apples does the store have left? 10 points

〈**Ans.**〉

9 Kevin has 26 red fish and 7 black fish in his pond. How many fish does he have in all? 10 points

〈**Ans.**〉

10 In the kennel, there are 24 cats and 9 dogs. How many more cats are there than dogs? 10 points

〈**Ans.**〉

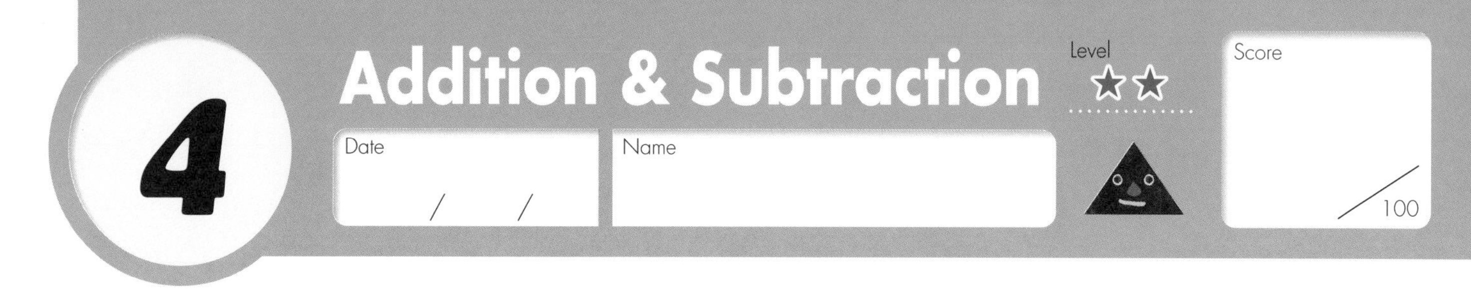

1. Allison made 15 cookies yesterday and 17 more today. How many cookies did she make altogether? 10 points

〈Ans.〉

2. Brandon had 26 dimes, and then used 17 dimes to buy some candy. How many dimes does he have left? 10 points

〈Ans.〉

3. At the pond, 24 ducks are swimming in the pond, and 16 are sitting on the bank. How many ducks are there in all? 10 points

〈Ans.〉

4. Uncle Brown keeps 24 cows and 18 horses at his farm. How many more cows than horses does he have? 10 points

〈Ans.〉

5. Teacher has 38 people in class and she gave each person one pencil. She still has 14 pencils left. How many pencils are there in all? 10 points

〈Ans.〉

6 In class today, we had 35 notebooks and 40 children. If each child gets one notebook, how many more notebooks will we need? 10 points

〈Ans.〉

7 You gave a sticker each to 18 people at school. If you have 27 stickers left, how many stickers did you have in all? 10 points

〈Ans.〉

8 Penny wants to give a tangerine to everyone at her party. If she has 28 tangerines, and 36 children are coming to her party, how many more tangerines does she need? 10 points

〈Ans.〉

9 Ron is working in the hen house on Monday. He has already packed 46 eggs and still has 14 left to pack. How many eggs are there altogether? 10 points

〈Ans.〉

10 On Tuesday, Ron is in the hen house. He has 32 eggs to pack. He has already put 18 in a box. How many eggs does he still need to pack? 10 points

〈Ans.〉

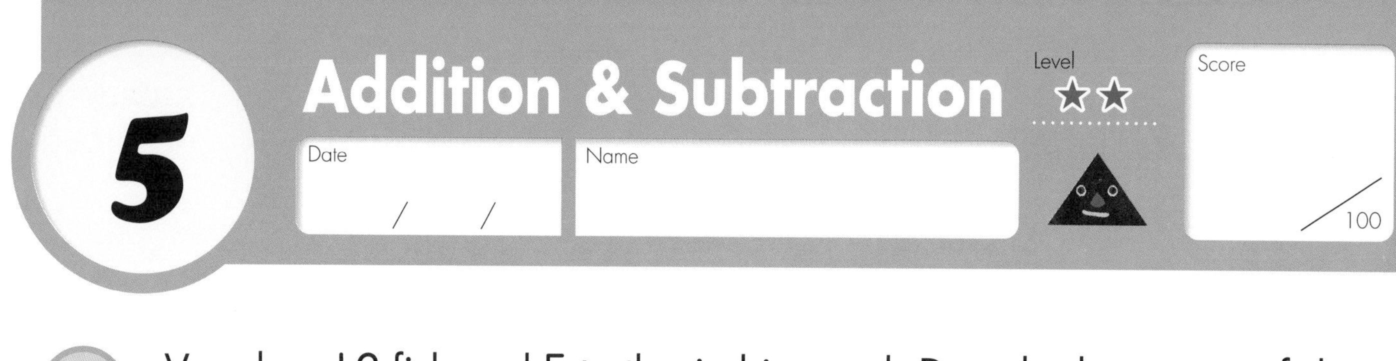

1 Vern has 12 fish and 5 turtles in his pond. Does he have more fish or turtles? How many more? 10 points

〈**Ans.**〉 He has ☐ more ☐.

2 Father has 7 apples and 12 pears in the kitchen. Does he have fewer apples or pears? How many fewer? 10 points

〈**Ans.**〉 He has ☐ fewer ☐.

3 The zoo has 18 raccoons and 25 foxes. Does the zoo have more raccoons or foxes? How many more? 10 points

〈**Ans.**〉 The zoo has ☐ more ☐.

4 Maria and Mike jumped rope. Maria jumped 62 times and Mike jumped 45 times. Who jumped more times? How many more? 10 points

〈**Ans.**〉 ☐ jumped ☐ more times.

5 We made 80 red paper airplanes and 74 white paper airplanes in class today. Did we make fewer red airplanes or white airplanes? How many fewer? 10 points

〈**Ans.**〉 We made ☐ fewer ☐ airplanes.

6 120 boys and 98 girls were in the gym today watching the game. Were there fewer boys or girls? How many fewer? 10 points

〈**Ans.**〉

7 At home, I have 145 comic books and 87 storybooks. Do I have fewer comic books or storybooks? How many fewer? 10 points

〈**Ans.**〉

8 Julie took a ferry to the island. On the ferry there were 108 men and 96 women. Were there more men or women on board? How many more? 10 points

〈**Ans.**〉

9 The castle had a large garden with 120 red roses and 95 white roses. Did they have more red or white roses? How many more? 10 points

〈**Ans.**〉

10 Ella folded 114 paper cranes, and Tom folded 98. Who folded fewer cranes? How many fewer? 10 points

〈**Ans.**〉

Now let's try something different!

6 Length

Date / /

Name

Level ☆☆

Score /100

1 You have two pieces of tape as seen on the right. If you connect both pieces together end-to-end, how long is your new piece of tape?

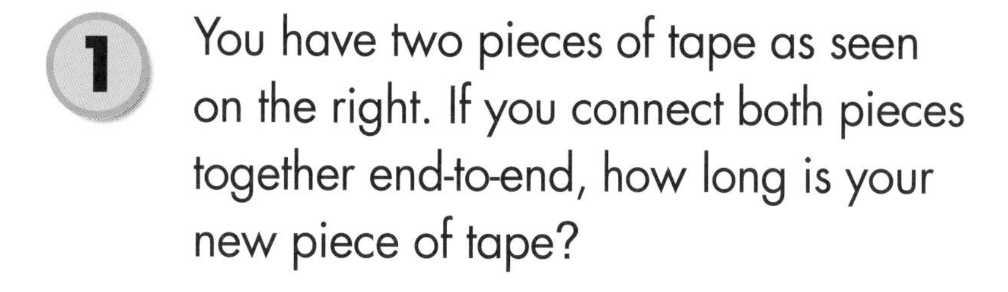

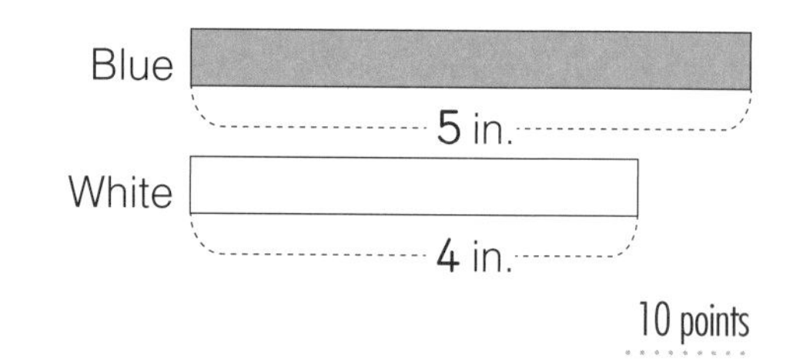

10 points

4 in. + 5 in. = 9 in.

〈**Ans.**〉 9 in.

2 In class, the teacher asks you to connect 6 inches of tape to 8 inches of tape end-to-end. How long is your new piece of tape?

10 points

6 in. + 8 in. =

〈**Ans.**〉

3 Kate knitted 16 inches of her scarf and wants to knit 14 inches more today. How long will the scarf be at the end of the day?

10 points

〈**Ans.**〉

4 A 5-foot stick and a 7-foot stick are on the ground. If you put them together, how long will the new stick be?

10 points

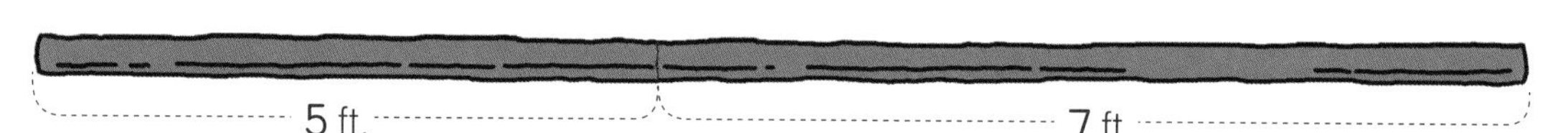

5 ft. + 7 ft. = 12 ft.

〈**Ans.**〉 12 ft.

5 What is the length from the left flag to the right flag shown below?

10 points

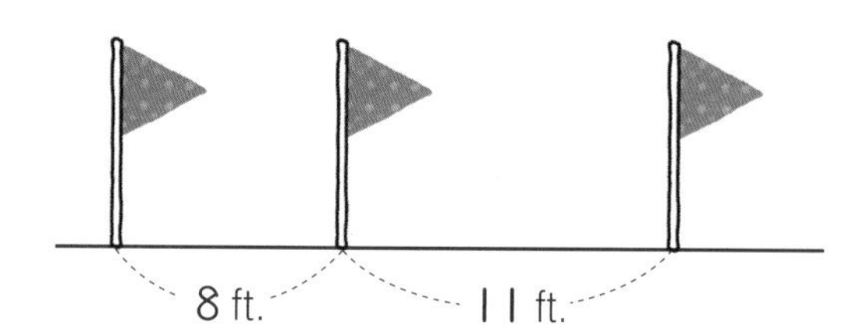

〈**Ans.**〉

6 The teacher has a 12-inch pencil and a 16-inch pencil. How much longer is the 16-inch pencil? 10 points

16 in. − 12 in. = 4 in.

〈**Ans.**〉 4 in.

7 Julian has a poster that is 18 inches long and 11 inches wide. How much longer is his poster than it is wide? 10 points

18 in. − 11 in. =

〈**Ans.**〉

8 You used 13 inches of wool out of 25 inches. How long is the remaining piece of wool? 10 points

〈**Ans.**〉

9 The scout master has an 8-foot rope and a 12-foot rope. What is the difference in length between the two ropes? 10 points

12 ft. − 8 ft. =

〈**Ans.**〉

10 Our gym is 42 feet long. The width of our gym is 14 feet less than the length. How wide is our gym? 10 points

〈**Ans.**〉

This is tough. Let's get some more practice in!

1 I foot 2 inches of tape and 3 inches of tape are joined together. How long is the new piece of tape? 10 points

I ft. 2 in. + 3 in. = I ft. 5 in.

〈**Ans.**〉

2 2 feet 5 inches of rope and 5 inches of rope are joined together. How long is the new piece of rope? 10 points

〈**Ans.**〉

3 I foot 7 inches of tape and 4 inches of tape are joined together. How long is the new piece of tape? 10 points

〈**Ans.**〉

4 A 4-foot 7-inch stick and a 4-foot stick are joined together as shown below. How long is the new stick? 10 points

4 ft. 7 in. 4 ft.

〈**Ans.**〉

5 A 2-foot 7-inch piece of ribbon and a 5-foot piece of ribbon are joined together as shown here. How long is the new ribbon? 10 points

〈**Ans.**〉

6 1 foot 10 inches of tape and 1 foot 1 inch of tape are connected together. How long is the new piece of tape? 10 points

〈**Ans.**〉

7 2 feet 8 inches of rope and 1 foot 2 inches of rope are joined together at the ends. How long is the new piece of rope? 10 points

〈**Ans.**〉

8 3 feet 5 inches of string and 1 foot 6 inches of string are connected together. How long is the new piece of string? 10 points

〈**Ans.**〉

9 A 2-foot 3-inch stick and a 1-foot 8-inch stick are joined together. How long is the new stick? 10 points

〈**Ans.**〉

10 A 5-foot 7-inch board and a 4-foot 1-inch board are joined together. How long is the new board? 10 points

〈**Ans.**〉

If you have trouble adding some measurements, try taking out some tape and a ruler!

1 Anna has 1 foot 11 inches of yellow tape. Raul has 10 inches of red tape. How much longer is Anna's tape? 10 points

〈Ans.〉

2 In class, we have 2 feet 9 inches of red tape left. We also have 5 inches of blue tape left. How much more red tape than blue tape do we have left? 10 points

〈Ans.〉

3 You used 8 inches of string out of the 2 feet 10 inches that we had. How much string is left now? 10 points

〈Ans.〉

4 At the gym, coach has a rope that is 8 feet 9 inches long, and a rope that is 5 feet long. What is the difference in length between the two ropes? 10 points

〈Ans.〉

5 Our gym is 42 feet long. The width is 16 feet less than the length. How wide is the gym? 10 points

〈Ans.〉

6 Father has 3 feet 9 inches of blue tape and 2 feet 5 inches of green tape. How much longer is the longer piece of tape? 10 points

〈**Ans.**〉

7 Holly's bathtub is 4 feet 10 inches long and 2 feet 7 inches wide. How long is the difference between the length and width of the bathtub? 10 points

〈**Ans.**〉

8 This classroom is 50 feet 8 inches long. The width is 15 feet 5 inches less than the length. How wide is this classroom? 10 points

〈**Ans.**〉

9 Vicki has a string that is 7 feet 11 inches long. Jamal has a string that is 4 feet 6 inches long. How much longer is the longer piece of string? 10 points

〈**Ans.**〉

10 You used 2 feet 3 inches of tape out of the 5 feet 9 inches you had. How long is the tape you have left? 10 points

〈**Ans.**〉

Are you getting the hang of this? Good!

1 You have two pieces of tape. The red tape is 7 centimeters long and the blue tape is 8 centimeters long. How long is the new piece of tape if you connect the two?

8 points

7 cm + 8 cm = 15 cm

〈Ans.〉

2 For each question below, write a formula in order to answer the question.

7 points per question

(1) I have a 15-centimeter string and a 12-centimeter string. If I connect them, how long will my new piece of string be?

〈Ans.〉

(2) You have a 4-meter and a 3-meter rope. If you connect them, how long will your new piece of rope be?

4 m + 3 m =

〈Ans.〉

(3) Father has a 6-centimeter piece of bamboo and a 5-centimeter piece of bamboo. If he glues them together, how long will his new piece of bamboo be?

6 cm + 5 cm =

〈Ans.〉

(4) The grass on your lawn is now 15 centimeters long. If it grows another 5 centimeters, how long will it be?

〈Ans.〉

(5) Wendy has a 1-meter rope and a 3-meter rope. If she connects them, how long is the new rope?

〈Ans.〉

(6) If you extend 11 centimeters of string by another 2 centimeters, how long will your new string be?

〈Ans.〉

3 The classroom had a 15-centimeter long glue stick, and then Tina used 7 centimeters of it. How long is the stick now? *8 points*

15 cm − 7 cm = 8 cm

〈**Ans.**〉 ____________

4 For each question below, write a formula in order to answer the question. *7 points per question*

(1) You had **20** centimeters of tape and then you used **8** centimeters of it. How long is the tape that is left?

20 cm − 8 cm =

〈**Ans.**〉 ____________

(2) Jill has a **4**-centimeter string. Mak has a **15**-centimeter string. What is the difference in length between the two strings?

〈**Ans.**〉 ____________

(3) On your father's boat, there is a **3**-meter rope and a **16**-meter rope. How much longer is the longer rope?

16 m − 3 m =

〈**Ans.**〉 ____________

(4) If you cut out **8** centimeters from a **54**-centimeter piece of string, how long is the string that is left?

〈**Ans.**〉 ____________

(5) The pool at the gym is **25** meters long and **10** meters wide. How much longer is the pool than it is wide?

〈**Ans.**〉 ____________

(6) At the dock, they cut a **15**-meter piece from a **70**-meter rope. How long is the rope that is left?

〈**Ans.**〉 ____________

Ropes and tape and sticks, oh my!

10 Length

Level ★★

Date / /

Name

Score /100

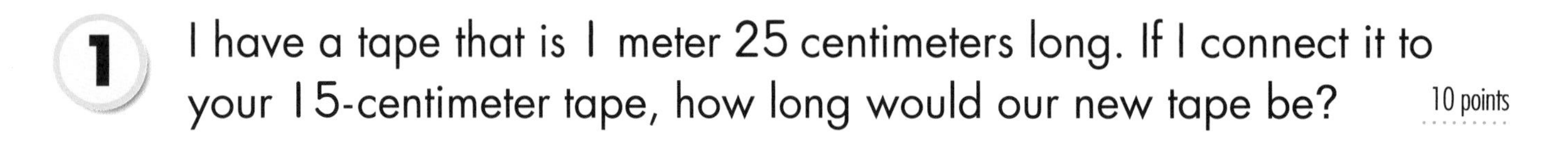

1 I have a tape that is 1 meter 25 centimeters long. If I connect it to your 15-centimeter tape, how long would our new tape be? 10 points

〈**Ans.**〉

2 In class, we had a tape that was 2 meters 47 centimeters long and a 22-centimeter tape. When we connected them, how long was the new tape? 10 points

〈**Ans.**〉

3 There is a piece of tape at home that is 1 meter 18 centimeters. If you take a 74-centimeter piece of tape home and connect it to that piece, how long is the new piece of tape? 10 points

〈**Ans.**〉

4 Jim is helping his father build a wall. If he connects a board that is 4 meters 27 centimeters long to another board that is 2 meters long, how long is the new board? 10 points

〈**Ans.**〉

5 Omar went and found the longest stick he could find, and it was 5 meters 84 centimeters long. He then added it to a stick Brian found that was 3 meters long. How long is the new stick? 10 points

〈**Ans.**〉

6 Ben wants to play tug of war with a really long rope. He got a rope that was 1 meter 46 centimeters long and connected it to a rope that was 2 meters 33 centimeters long. How long is the new rope? 10 points

〈**Ans.**〉

7 We found a rope that was 2 meters 13 centimeters long and connected it to a rope that was 1 meter 74 centimeters long. How long is our new rope? 10 points

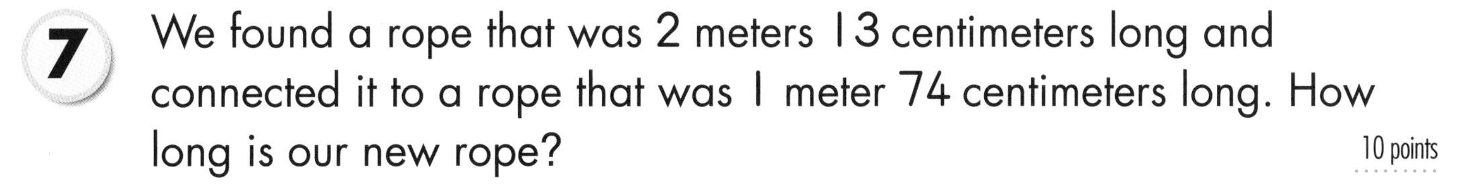

〈**Ans.**〉

8 Deb has a string that is 1 meter 38 centimeters long. I have a string that is 3 meters 41 centimeters long. How long would our string be if we connected them? 10 points

〈**Ans.**〉

9 During lunchtime, Sammy found two sticks. One was 2 meters 17 centimeters long, and another was 1 meter 48 centimeters long. How long would the new stick be if he glued them together? 10 points

〈**Ans.**〉

10 To make our ramp, we attached a board that was 1 meter 73 centimeters long to another board that was 1 meter 25 centimeters long. How long is our new board? 10 points

〈**Ans.**〉

You're cooking now!

1. Janelle has 1 meter 58 centimeters of white tape. She also has 1 meter 11 centimeters of pink tape. How much longer is the white tape? 10 points

〈Ans.〉

2. In class, we have a gold banner that is 2 meters 67 centimeters long. We also have a silver banner that is 1 meter 23 centimeters long. What is the difference in the length of the two banners? 10 points

〈Ans.〉

3. My mother really likes string. She has a string that is 6 meters 72 centimeters long, and another that is 3 meters 43 centimeters long. How much longer is the longer string? 10 points

〈Ans.〉

4. Benji found a stick that was 2 meters 84 centimeters long. He sawed it into two pieces, and one piece was 1 meter 31 centimeters long. How long was the other stick? 10 points

〈Ans.〉

5. Our backyard is 25 meters 65 centimeters long. The width of our backyard is 7 meters 35 centimeters less than the length. How wide is our backyard? 10 points

〈Ans.〉

6 Jennifer has a tape that is 3 meters 98 centimeters long. Libby's tape is 2 meters 19 centimeters long. How much longer is Jennifer's tape?

10 points

〈**Ans.**〉

7 The road by my house is 55 meters 75 centimeters long. The width of the road is 50 meters 50 centimeters shorter than the length. How wide is the road by my house?

10 points

〈**Ans.**〉

8 Grandmother's garden is 10 meters 55 centimeters long and 27 meters 65 centimeters wide. What is the difference between the width and the length of her garden?

10 points

〈**Ans.**〉

9 Tom was comparing his string, which was 5 meters 86 centimeters long, to Olivia's string, which was 2 meters 52 centimeters long. How much longer was Tom's string?

10 points

〈**Ans.**〉

10 You used 2 meters 54 centimeters of tape packing presents. We had 6 meters 73 centimeters when you started. How long is the tape now?

10 points

〈**Ans.**〉

Let's try something a little different!

1 We have 2 pounds of bananas and 5 pounds of grapes for the big fruit salad we are making for Sunday. How many pounds of fruit do we have altogether? 10 points

2 lb. + 5 lb. =

〈Ans.〉

2 Ken went shopping and put 1 pound of sugar and 14 pounds of coffee in his cart first. How much do the things in his cart weigh now? 10 points

〈Ans.〉

3 The cooks for the big holiday party got 25 pounds of ham and 12 pounds of bacon. How much does their meat weigh in all? 10 points

〈Ans.〉

4 The store owner has 74 pounds of tomatoes. If he bought 11 more pounds, how much would all of his tomatoes weigh together? 10 points

〈Ans.〉

5 Andy weighed 55 pounds last year. His weight increased 15 pounds this year. How much does he weigh now? 10 points

now
last year

〈Ans.〉

6 The grocery truck has 23 pounds of yellow onions and 9 pounds of red onions. What is the difference in weight between the two types of onions? 10 points

〈**Ans.**〉

7 Tammy's farm produced 87 pounds of watermelon and 42 pounds of melon this month. How much more watermelon than melon did she produce? 10 points

〈**Ans.**〉

8 Jen weighs 49 pounds, and Jane weighs 51 pounds. What is the difference in their weights? 10 points

〈**Ans.**〉

9 The cake shop had 87 pounds of sugar this morning. Since then, the chef has used 35 pounds of sugar. How much sugar is left? 10 points

〈**Ans.**〉

10 24 pounds out of the 53 pounds of ham at the meat shop were sold today. What is the weight of the remaining ham? 10 points

〈**Ans.**〉

Not so bad, right? Good job!

1 We bought 3 kilograms of apples and 2 kilograms of oranges today at the store. How much did the fruit weigh altogether? 10 points

3 kg + 2 kg =

〈**Ans.**〉

2 Rob is transporting 5 kilograms of sugar and 20 kilograms of flour to the bakery. How much weight is he carrying in all? 10 points

〈**Ans.**〉

3 Will loaded 27 kilograms of iron and 15 kilograms of copper into a truck. How much did the metal weigh in all? 10 points

〈**Ans.**〉

4 The kitchen at the cafeteria has 52 kilograms of potatoes. If the chef bought 24 more kilograms of potatoes, how much do all of his potatoes weigh in all? 10 points

〈**Ans.**〉

5 Tom shipped 31 kilograms of wheat yesterday, and then he shipped 55 kilograms of wheat today. How much did he ship in all? 10 points

〈**Ans.**〉

6 Dara bought 12 kilograms of apples while Greg bought 8 kilograms of apples. How much more do Dara's apples weigh? 10 points

〈Ans.〉

7 In the pantry, there are 34 kilograms of red beans and 15 kilograms of green beans. What is the difference in weight between the two types of beans? 10 points

〈Ans.〉

8 The cafeteria had 56 kilograms of flour this morning. If they used 35 kilograms of flour today, how much flour is left? 10 points

〈Ans.〉

9 John weighs 36 kilograms and Cindy weighs 27 kilograms. How much more does John weigh? 10 points

〈Ans.〉

10 Steve was making cement and used 24 kilograms of sand today. If he had 50 kilograms of sand this morning, how much sand does he have now? 10 points

〈Ans.〉

You've got it now!

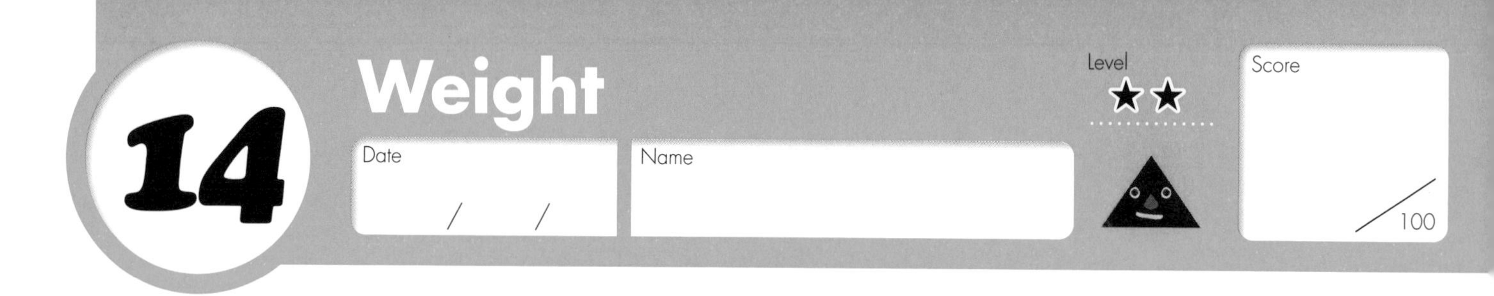

1 We bought 5 kilograms of pears and 7 kilograms of pineapples today at the store. How much did the fruit weigh altogether? 10 points

〈**Ans.**〉

2 The bakery had a 1-pound bag of sugar and a 10-pound bag of flour at the end of the day. Jim picked up both bags to move them to the storeroom. What did the two bags weigh in all? 10 points

〈**Ans.**〉

3 Sun's father weighed 57 kilograms three years ago. His weight increased 13 kilograms this year. How much does he weigh now? 10 points

〈**Ans.**〉

4 The store owner has 43 kilograms of corn. If he bought 11 more kilograms, how much would all of his corn weigh? 10 points

〈**Ans.**〉

5 Tom shipped 76 pounds of oranges yesterday, and then he shipped 51 pounds of oranges today. How much did he ship in all? 10 points

〈**Ans.**〉

6 Grandfather loves jam. He has 13 pounds of strawberry jam and 18 pounds of apple jam. What is the difference in weight between the two types of jam? 10 points

⟨Ans.⟩ ____________

7 Jennifer is moving to a new house and packing up. She has 44 kilograms of books and 28 kilograms magazines. How much more do her books weigh than her magazines? 10 points

⟨Ans.⟩ ____________

8 The deli had 50 pounds of salad this morning. They sold 27 pounds during the day today. How much salad do they have left? 10 points

⟨Ans.⟩ ____________

9 Karen's dog weighs 15 kilograms and Sophie's dog weighs 17 kilograms. How much more does Sophie's dog weigh? 10 points

⟨Ans.⟩ ____________

10 George weighed 75 pounds last year. This year, he got sick and lost 7 pounds. How much does he weigh now? 10 points

⟨Ans.⟩ ____________

Okay, let's switch it up a little!

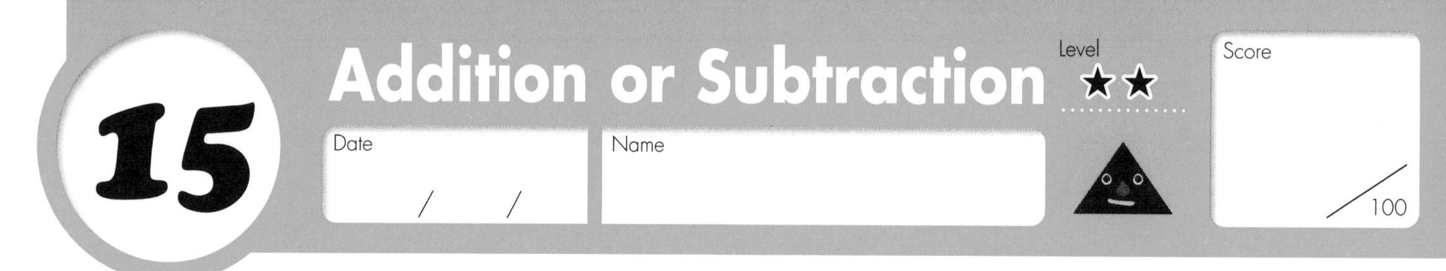

1 A crowd of children is playing at the park. 4 children join them, and then a little later 5 more come. How many children joined the crowd?

8 points

〈**Ans.**〉

2 Some ducks are swimming around the pond. 6 ducks get in the water and start swimming, and then 3 more join in. How many additional ducks are swimming around the pond now?

8 points

〈**Ans.**〉

3 The chickens were eating their feed when 3 got full and left. Then another 5 left the cage. How many chickens left the cage?

8 points

〈**Ans.**〉

4 A bus pulled up to a station, and 4 people got off the bus. 9 people got on. How many additional people are on the bus now?

8 points

〈**Ans.**〉

5 Another bus pulled up to the same station, and 8 people got off. Only 3 people got on. How many fewer people are on the bus now?

8 points

〈**Ans.**〉

6 16 pigeons were eating bread crumbs at the park. 2 of them flew away, and then 3 of them flew away after that.

10 points per question

(1) How many pigeons flew away in all?

2 + 3 =

〈Ans.〉

(2) How many pigeons stayed behind?

16 − 5 =

〈Ans.〉

7 You have 18 apples at home because you love apples. You ate 5 yesterday and 3 today.

10 points per question

(1) How many apples did you eat?

〈Ans.〉

(2) How many apples do you still have?

〈Ans.〉

8 18 children were playing tag in the park. 4 children joined in, and then 3 more children joined in.

10 points per question

(1) How many children joined in?

〈Ans.〉

(2) How many children are playing tag in all?

〈Ans.〉

One step at a time! Well done.

Addition or Subtraction

Level ★★

Date / /

Name

Score /100

1 There were 20 eggs in the kitchen. You ate 4 eggs yesterday and then 5 eggs this morning.

10 points per question

(1) How many eggs did you eat?

4 + 5 =

⟨**Ans.**⟩

(2) How many eggs are left? Use the formula below.

[20] − ([4 + 5]) = [11]

⟨**Ans.**⟩

2 You bought a pencil and a notebook with the price tags shown below. You had 100¢. How much change did you get? Use the formula below.

10 points

[100] − ([30 + 60]) = [ ]

⟨**Ans.**⟩

3 There are 40 stickers in your sticker book. You used 15 yesterday and 9 today. How many stickers are left?

10 points

40 − (15 + 9) =

⟨**Ans.**⟩

4 There were 30 people on the bus. 8 people got off at the train station, and 6 people got off at the hospital. How many people are still in the bus? 12 points

〈**Ans.**〉 ____________

5 19 sparrows were sitting on the fence. 5 of them flew away when Ben sat nearby. 4 of them flew away when he got up and walked away. How many sparrows are still on Ben's fence? 12 points

〈**Ans.**〉 ____________

6 I had 35 baseball cards in my backpack this morning. I gave 10 to my friend Bill, and then I lost 8 today. How many cards remain in my backpack? 12 points

〈**Ans.**〉 ____________

7 You had 20 chocolate bars. You gave 6 to your brother and also gave 6 to your sister. How many bars do you have left? 12 points

〈**Ans.**〉 ____________

8 Vicky had 50 strawberries in the kitchen. She ate 9 yesterday and 8 today. How many strawberries does she have left? 12 points

〈**Ans.**〉 ____________

17 Addition or Subtraction

Level ★★

Date / /

Name

Score /100

1 Carol used 3 sheets of colored paper. There are still 5 sheets of colored paper remaining. How many sheets of paper did she have at the beginning? 10 points

5 + 3 = 8

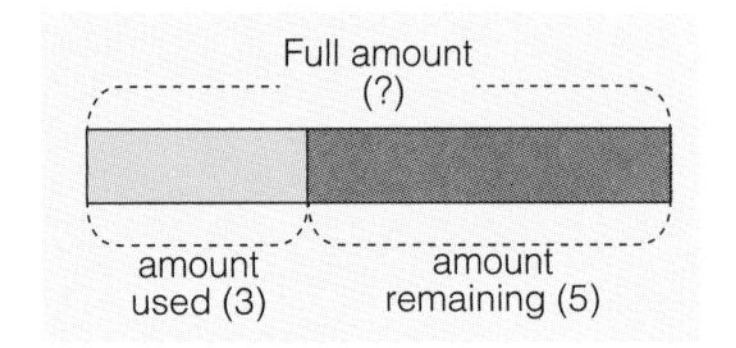

〈Ans.〉 ______________

2 Monica's roses are blooming. She cut 4 flowers. There are still 6 flowers remaining. How many flowers did she have before she cut any? 10 points

6 + =

〈Ans.〉 ______________

3 You bought some apples at the store yesterday. Today, you ate 2 of them. You still have 5 apples remaining. How many apples did you buy? 10 points

5 + =

〈Ans.〉 ______________

4 James ate 5 oranges from the basket. He still has 8 left. How many oranges were in the basket before he ate any? 10 points

〈Ans.〉 ______________

5 The students in class A used 38 sheets of paper today. There are still 20 sheets of paper remaining. How many sheets of paper did they have before? 10 points

〈Ans.〉 ______________

6 Bill is reading a book about space. He read 85 pages this week. He has 75 pages left. How many pages does the book have? 10 points

〈**Ans.**〉

7 I bought lollipops for my school. I gave away 35 so far, and I have 95 more to give. How many lollipops did I buy? 10 points

〈**Ans.**〉

8 You used 25 stickers in your sticker book. You still have 75 left. How many stickers were in your sticker book at first? 10 points

〈**Ans.**〉

9 Lora used 60 centimeters of tape. Now, the tape is 2 meters 30 centimeters long. How long was the tape before she used it? 10 points

〈**Ans.**〉

10 Peter shipped 85 bags by truck. He still has 35 bags. How many bags did he have before he shipped any? 10 points

〈**Ans.**〉

This looks hard,
but it's not so bad, right?

18 Addition or Subtraction

Level ★★

Date / /

Name

Score /100

1 You had 8 colored sheets of paper. You used some sheets, and now you have 5 sheets. How many sheets of paper did you use? 10 points

8 − 5 = 3

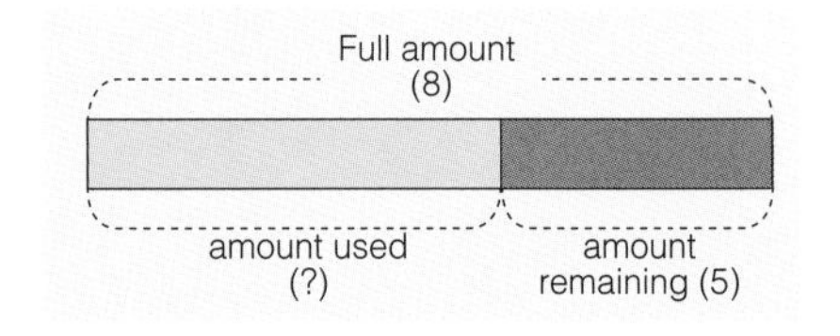

⟨Ans.⟩ ______

2 10 sparrows were on the wire. Some sparrows flew away, and there were 6 sparrows remaining on the wire. How many sparrows flew away? 10 points

10 − =

⟨Ans.⟩ ______

3 The pastor had 7 candles. He used some candles for the service, and now he has 2 candles. How many candles did the pastor use? 10 points

7 − =

⟨Ans.⟩ ______

4 You had 9 apples. You gave your neighbor some apples, and now you have 5 apples. How many apples did you give away? 10 points

⟨Ans.⟩ ______

5 There were 12 fish in the pond yesterday. Today, somebody caught some of the fish. Now there are 10 fish in the pond. How many fish were caught today? 10 points

⟨Ans.⟩ ______

6 Grandmother had 14 postcards. She sent some off today, and now she has 8. How many postcards did Grandmother use? 10 points

⟨**Ans.**⟩

7 There were 24 pigeons in the park, until some people walked by and scared some of them away. Now there are only 8 pigeons left. How many pigeons flew away? 10 points

⟨**Ans.**⟩

8 You had 150 blueberries and ate some yesterday. Today, you have 80. How many blueberries did you eat yesterday? 10 points

⟨**Ans.**⟩

9 Anna had 30 centimeters of tape until she used some for a craft she was making. Now she has 7 centimeters of tape. How much did she use? 10 points

⟨**Ans.**⟩

10 The cafeteria had 100 eggs this morning. After breakfast, the cafeteria now has 64 eggs. How many eggs did the cafeteria use during breakfast this morning? 10 points

⟨**Ans.**⟩

You're getting better all the time!

19 Addition or Subtraction

Level 

Date / / Name

Score /100

1 You had 3 red sheets of paper, and then you went and got some blue sheets of paper. Altogether, you have 8 sheets of paper now. How many blue sheets of paper did you get?

10 points

8 − 3 = 5

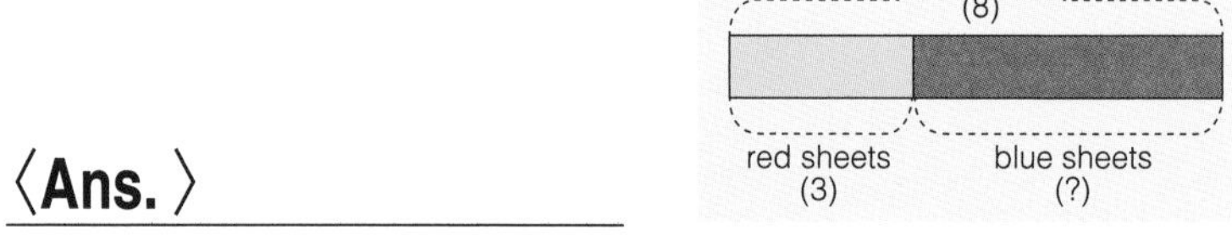

⟨Ans.⟩ ____________

2 6 people were playing ball when some more people joined in. Now there are 10 people playing ball. How many people just joined the game?

10 points

10 − =

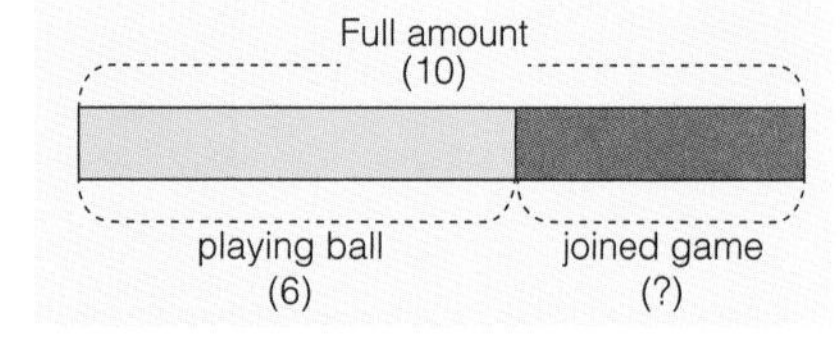

⟨Ans.⟩ ____________

3 There were 4 ships on the dock this morning. Many ships returned during the day, and now there are 12 ships. How many ships returned today?

10 points

12 − =

⟨Ans.⟩ ____________

4 Yannick had 10 dimes, but then he was given some more for his birthday. Now he has 30 dimes. How many dimes did he get for his birthday?

10 points

⟨Ans.⟩ ____________

5 Reina made 28 cookies yesterday. Today, she baked more cookies and now there are 64 cookies in all. How many cookies did she bake today?

10 points

⟨Ans.⟩ ____________

6 Yesterday, Jack read 35 pages in his book. Today, he read some more. In all, he read 72 pages. How many pages did he read today? 10 points

〈Ans.〉

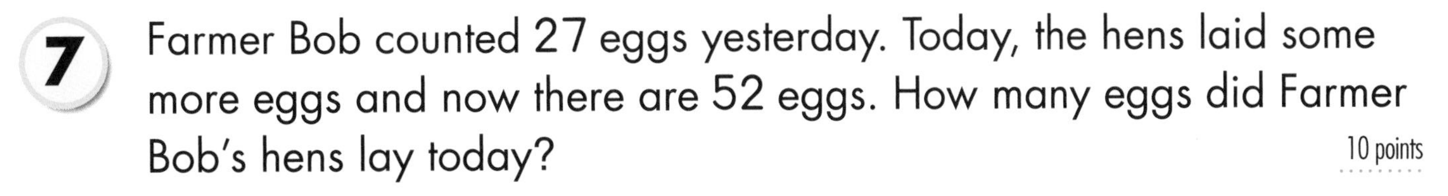

7 Farmer Bob counted 27 eggs yesterday. Today, the hens laid some more eggs and now there are 52 eggs. How many eggs did Farmer Bob's hens lay today? 10 points

〈Ans.〉

8 Lilly has 65 building blocks. This morning, her mother gave her some for her birthday. Now she has 125 blocks. How many blocks did her mother give her today? 10 points

〈Ans.〉

9 Amy's family restaurant had 42 apples. Her grandmother came by today and delivered some more apples. If they have 60 apples now, how many apples did her grandmother bring? 10 points

〈Ans.〉

10 Yesterday, Rima picked up 87 acorns. Today, she picked up some more and now she has 141 acorns. How many acorns did Rima pick up today? 10 points

〈Ans.〉

Phew! Take a break if you are tired.
This is getting tough.

20 Addition or Subtraction

Level ★★

Date / /

Name

Score

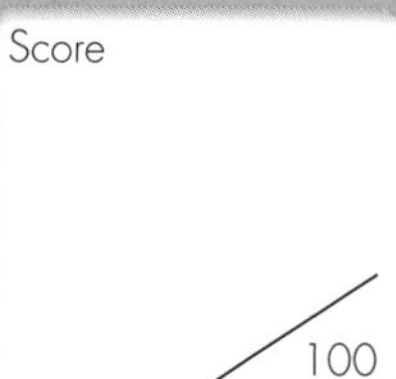

1 My father gave me 4 coins. Now I have 9 coins. How many coins did I have before?

10 points

9 − 4 = 5

Full amount (9)

coins I had before (?)

coins given (4)

〈**Ans.**〉

2 In class, we have some tadpoles in a tank. Today, you put 3 tadpoles in the tank, and now there are 10 tadpoles. How many tadpoles were in the tank yesterday?

10 points

10 − =

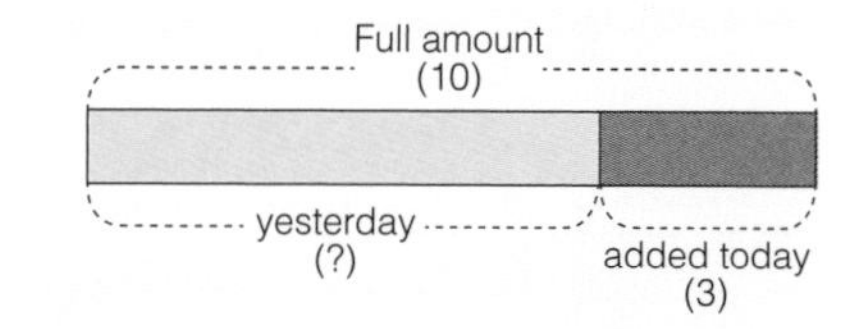

〈**Ans.**〉

3 Today, Sally folded 3 paper cranes. Now there are 8 paper cranes in class. How many cranes were there in class yesterday?

10 points

8 − =

〈**Ans.**〉

4 Jessica got 40¢ from her mother, and now she has 100¢. How many cents did she have before?

10 points

〈**Ans.**〉

5 Julia brought 5 goldfish home from the store today. Now she has 23 goldfish at home. How many goldfish did she have before?

10 points

〈**Ans.**〉

6 9 pigeons landed by the lady with the breadcrumbs and now there are 25 pigeons over there. How many pigeons were there before?

10 points

〈Ans.〉

7 16 children arrived at the park. 38 children are now playing at the park. How many children were playing before?

10 points

〈Ans.〉

8 Sara read 37 pages in her book today. In all, she has read 65 pages. How many pages did she read before today?

10 points

〈Ans.〉

9 Mary got 25 stickers from her sister. Now she has 72 stickers. How many stickers did she have before she got some from her sister?

10 points

〈Ans.〉

10 Bill sold his old bicycle for $70. Now, he has $130 saved. How much did he have before he sold his bicycle?

10 points

〈Ans.〉

You're doing really awesome!

1. Brian bought some apples yesterday. Today, he ate 7 apples with his brother. Now there are 9 apples left. How many apples did Brian buy yesterday?
10 points

⟨**Ans.**⟩

2. Maria used 12 sheets of paper. Now there are 20 sheets left. How many sheets of paper did she have before?
10 points

⟨**Ans.**⟩

3. 16 ducks were swimming around the pond at the park. Some ducks got out and now there are 7 ducks in the pond. How many ducks got out?
10 points

⟨**Ans.**⟩

4. There were 45 hair clips in the bathroom upstairs. Sophie and her sisters used some this week, and now there are only 6 clips upstairs. How many clips did Sophie and her sisters use?
10 points

⟨**Ans.**⟩

5. In the craft room, there was a 16-meter tape. After craft class today, the tape is only 4 meters long. How much tape did class use today?
10 points

⟨**Ans.**⟩

6 There are 9 red flowers in Irene's garden. She brought home some yellow flowers today and now there are 17 flowers in all. How many yellow flowers did she bring home today? 10 points

⟨**Ans.**⟩

7 You had 16 baseball cards. Your sister gave you some baseball cards, and now you have 30. How many cards did your sister give you? 10 points

⟨**Ans.**⟩

8 Today 6 chicks hatched at the farm, and now there are 25 chicks running around. How many chicks were there yesterday? 10 points

⟨**Ans.**⟩

9 Anna got 35 stickers from her brother. Now she has 82 total stickers. How many stickers did she have before her brother gave her some? 10 points

⟨**Ans.**⟩

10 Marty said that if he had 45 more chocolate chips, he could bake a cookie cake that needs 120 chocolate chips. How many chocolate chips does he have now? 10 points

⟨**Ans.**⟩

22 Addition or Subtraction

Level 

Date / /

Name

Score /100

1 At the market, an apple costs 90¢. The price of an apple is 50¢ higher than the price of an orange. How much does the orange cost?

10 points

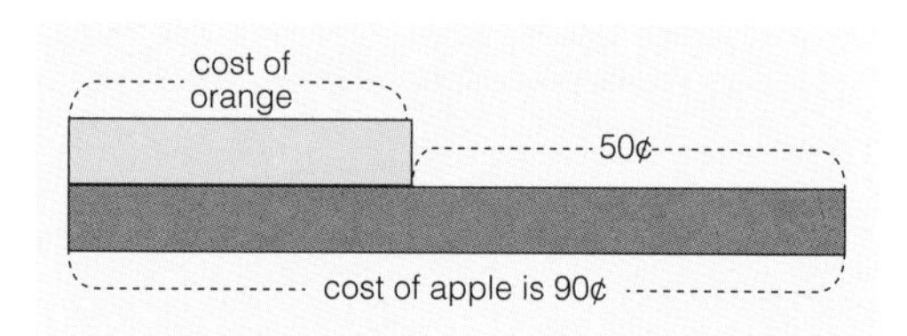

⟨**Ans.**⟩ ____________

2 There is a chocolate bar at the store that costs 90¢. The cost of that chocolate bar is 40¢ higher than the cost of a pack of gum. How much is the pack of gum?

10 points

90 − =

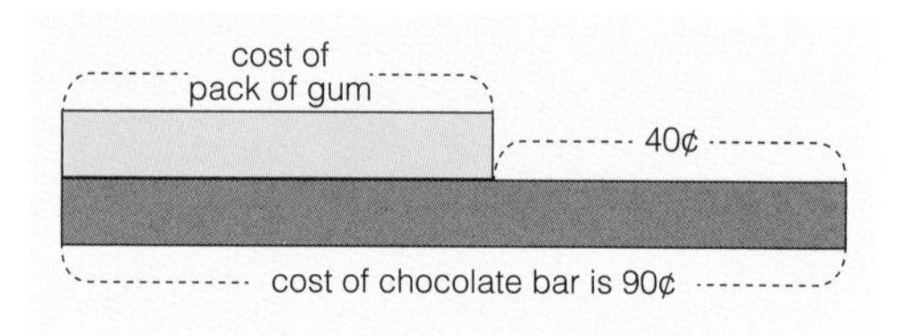

⟨**Ans.**⟩ ____________

3 Lucy bought a notebook for school for 80¢. Her notebook cost 50¢ more than a pencil. How much does the pencil cost?

10 points

80 − =

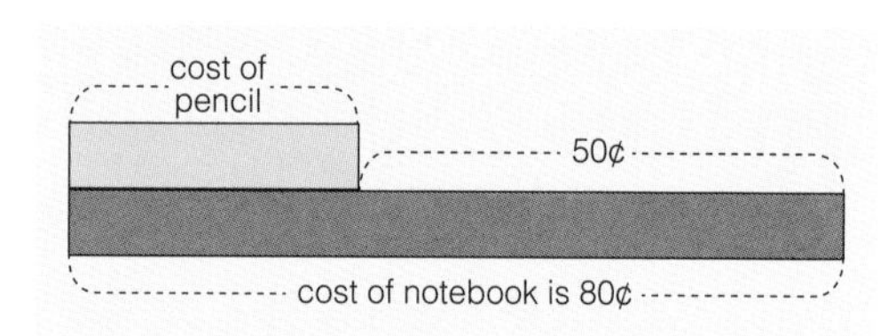

⟨**Ans.**⟩ ____________

4 The ballpoint pen costs 30¢ more than an eraser. If the ballpoint pen costs 70¢, how much is the eraser?

10 points

⟨**Ans.**⟩ ____________

5 Harry bought a pencil for school. The pencil cost 25¢ more than the eraser. If the pencil cost 80¢, how much was the eraser? 10 points

cost of eraser
25¢
cost of pencil is 80¢

〈**Ans.**〉

6 Ted has 62 pennies. He has 15 more pennies than Sally. How many pennies does Sally have? 10 points

〈**Ans.**〉

7 We have 95 centimeters of blue ribbon for the dress. The blue ribbon is 28 centimeters longer than the red ribbon. How long is the red ribbon? 10 points

〈**Ans.**〉

8 The necklace has 60 more beads on it than the bracelet. The necklace has 140 beads. How many beads are on the bracelet? 15 points

〈**Ans.**〉

9 There are 96 more sheets of white paper than graph paper in class. If there are 124 sheets of white paper, how many sheets of graph paper are there? 15 points

〈**Ans.**〉

I know it's tough, but you can do it.

23 Addition or Subtraction

Level ★★

Date / /

Name

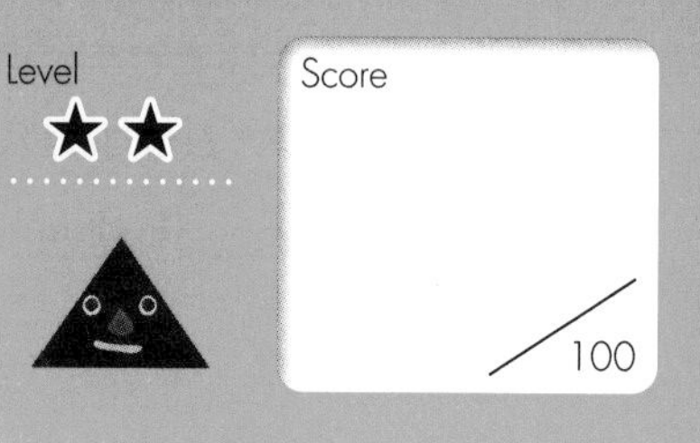

1 Allison is looking at a hair clip that costs 40¢. The price of the hair clip is 30¢ less than a comb. How much is the comb?

10 points

40 + 30 = 70

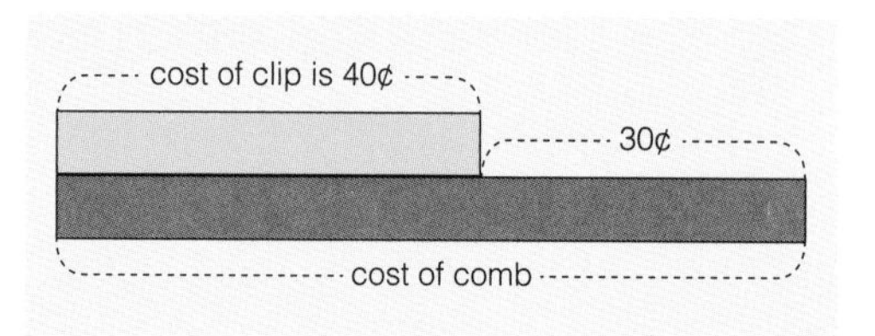

〈Ans.〉

2 Justin has 70 coins. He has 20 fewer coins than his brother. How many coins does Justin's brother have?

10 points

70 + =

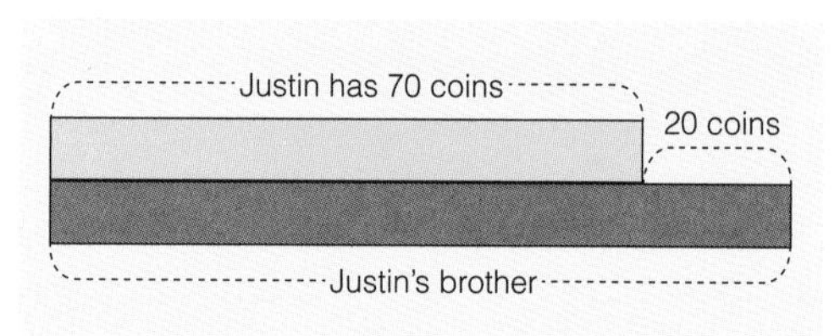

〈Ans.〉

3 Ella bought a donut that was 45¢ less expensive than a brownie. The price of the donut was 30¢. How much did the brownie cost?

10 points

30 + =

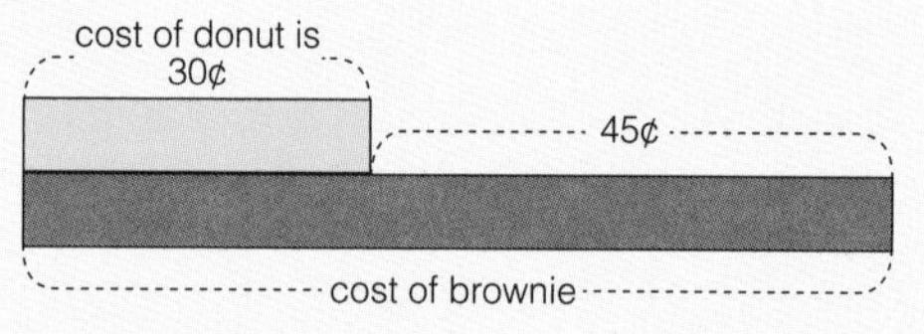

〈Ans.〉

4 At the corner store, an orange costs 55¢ less than an apple. If the orange costs 35¢, how much does the apple cost?

10 points

〈Ans.〉

5 For crafts class, our teacher brought some red and blue sheets of paper. There are 26 fewer blue sheets than red sheets. If there are 48 blue sheets, how many red sheets are there? 10 points

48 blue sheets
26 sheets
red sheets

〈Ans.〉

6 Sam has 35 stickers, and he has 17 fewer than his sister. How many stickers does his sister have? 10 points

〈Ans.〉

7 While they were shopping, Helen bought 60 markers. If she bought 90 fewer markers than her sister, how many did her sister buy? 10 points

〈Ans.〉

8 Mary bought an animal puzzle that had 65 pieces less than the flower puzzle. The animal puzzle had 85 pieces. How many pieces did the flower puzzle have? 15 points

〈Ans.〉

9 We are trying to choose between some wire and some string in order to wrap our present. We know the wire is 58 centimeters shorter than the string, and that the wire is 27 centimeters long. How long is the string? 15 points

〈Ans.〉

Good job!

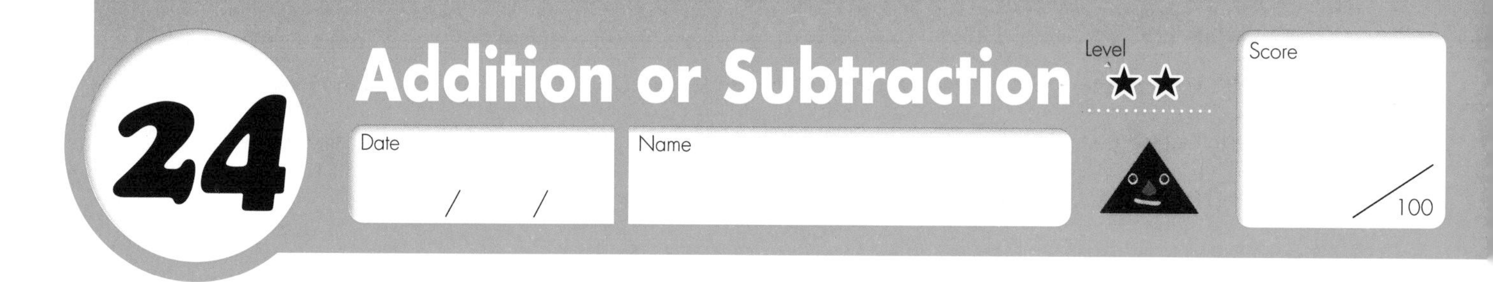

Addition or Subtraction

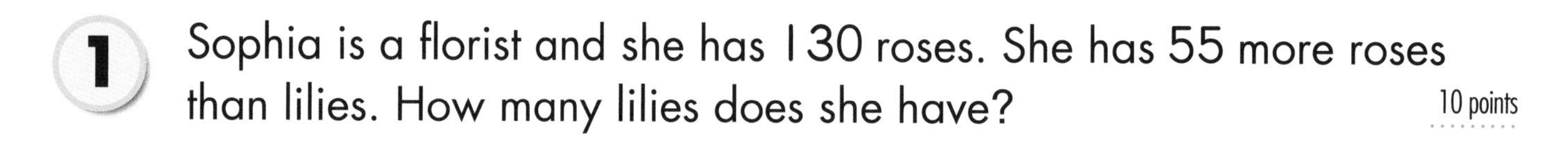

1 Sophia is a florist and she has 130 roses. She has 55 more roses than lilies. How many lilies does she have? 10 points

〈Ans.〉 ____________

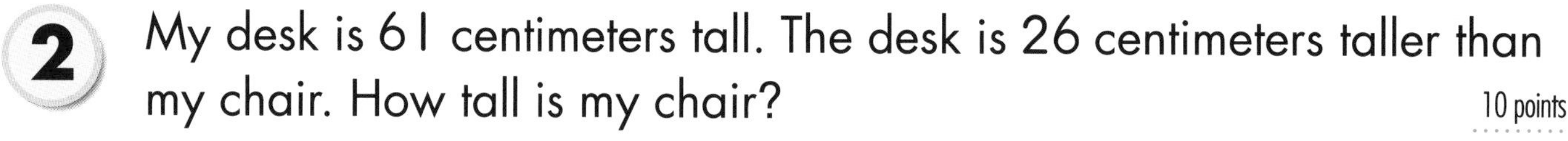

2 My desk is 61 centimeters tall. The desk is 26 centimeters taller than my chair. How tall is my chair? 10 points

〈Ans.〉 ____________

3 The toy store stocks 60 fewer marbles than jump ropes. If they stock 85 marbles, how many jump ropes will they have? 10 points

〈Ans.〉 ____________

4 Mary is reading a book about fairies. She read 37 pages today, which was 14 pages less than she read yesterday. How many pages did she read yesterday? 10 points

〈Ans.〉 ____________

5 Mother is 32 years younger than Grandfather and she is now 36 years old. How old is Grandfather? 10 points

〈Ans.〉 ____________

6 For the winter play, the school gathered in the gym. There were 102 children, which was 16 more than the number of adults. How many adults came to the play?

10 points

⟨**Ans.**⟩

7 We have 67 sheets of white paper in class. We have 28 fewer sheets of white paper than graph paper. How many sheets of graph paper do we have?

10 points

⟨**Ans.**⟩

8 Hanna bought a honeydew melon that had 80 seeds in it. If the honeydew melon had 25 less seeds than the watermelon, how many seeds did the watermelon have?

10 points

⟨**Ans.**⟩

9 Ted harvested 69 peaches from two trees on his farm. He harvested 14 more peaches than melons. How many melons did he harvest?

10 points

⟨**Ans.**⟩

10 Anne picked up 75 acorns, but Mike picked up 8 more acorns than she did. How many acorns did Mike pick up?

10 points

⟨**Ans.**⟩

Now it's time to mix it up!

25 Mixed Calculations

Level ★★★

Date / / Name

Score /100

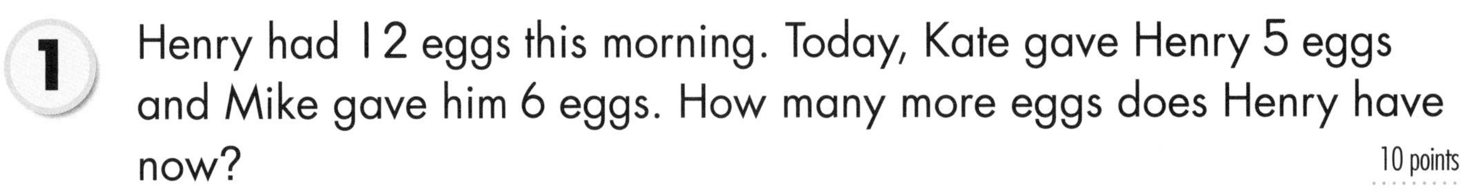

1 Henry had 12 eggs this morning. Today, Kate gave Henry 5 eggs and Mike gave him 6 eggs. How many more eggs does Henry have now? 10 points

5 + 6 = 11

⟨**Ans.**⟩ ______

2 Brenda had 15 pencils. Today, her sister gave her 6 more and her mother gave her 12 more. How many more pencils does Brenda have now? 10 points

☐ + ☐ = ☐

⟨**Ans.**⟩ ______

3 I had 80 centimeters of string before I started wrapping presents. I used 25 centimeters yesterday, and 18 centimeters today. How much string did I use while wrapping presents? 10 points

☐ cm + ☐ cm = ☐ cm

⟨**Ans.**⟩ ______

4 You had to write 46 thank-you notes. Yesterday, you wrote 18 notes. Today, you wrote 15 notes. How many notes did you write? 10 points

⟨**Ans.**⟩ ______

5 On Monday, Bobby's class ate 15 oranges. Tuesday, his class ate 17 oranges. Now, there are 18 left. How many fewer oranges does Bobby's class have than it had at the beginning of the week? 10 points

⟨**Ans.**⟩ ______

6 Dave had 100 stickers before his birthday. On his birthday, he got 70 stickers from his father and 50 stickers from his grandmother. How many more stickers does he have now? 10 points

〈Ans.〉

7 The ice cream store had 25 ice cream bars in their freezer. 75 ice cream bars were delivered yesterday and 65 ice cream bars were delivered today. How many more bars do they have now? 10 points

〈Ans.〉

8 The local store had 182 packs of gum in the back. Yesterday, they moved 85 packs of gum to the front, and today, they moved 43 packs of gum to the front. How many fewer packs do they have in the back? 10 points

〈Ans.〉

9 You got 1 box of oranges with 124 oranges in it. You gave 80 oranges to your friend's family, and you ate 32 oranges with your family. How many fewer oranges do you have? 10 points

〈Ans.〉

10 Sam is 126 centimeters tall. He found a 75-centimeter platform and put a 60-centimeter chair on it. Once he stood on the chair, how much taller was he? 10 points

〈Ans.〉

Pay close attention to what the question is asking for! Well done.

26 Mixed Calculations

Level ★★★

Date / /

Name

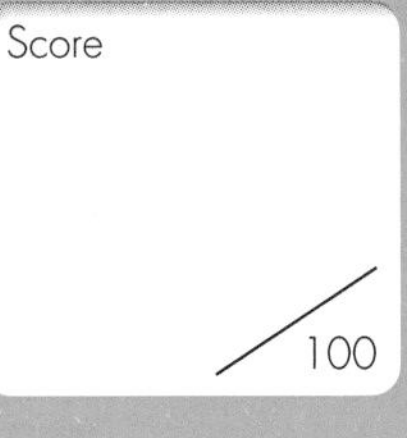

Score /100

1 Mary had 30 sheets of colored paper. Ellen had 20. They each used 5 sheets to do their crafts.

10 points per question

(1) How many sheets does each girl have left?

⟨Mary⟩ $30-5=25$

⟨Ans.⟩ 25 sheets

⟨Ellen⟩ $20-5=15$

⟨Ans.⟩ 15 sheets

(2) How many more sheets does Mary have left than Ellen?

$25-15=10$

⟨Ans.⟩ 10 sheets

(3) When they started, how many more sheets did Mary have than Ellen?

$30-20=10$

⟨Ans.⟩ 10 sheets

2 Ava and Ted each got money from their parents. Ava got 80¢ and Ted got 50¢. They each bought one 40¢ eraser.

10 points per question

(1) How much money does each have left?

⟨Ava⟩

⟨Ans.⟩

⟨Ted⟩

⟨Ans.⟩

(2) Now, how much more money does Ava have than Ted?

⟨Ans.⟩

(3) Before they bought the eraser, how much more money did Ava have than Ted?

⟨Ans.⟩

3 Lisa had 100¢ and David had 80¢. They each bought a 30¢ pack of gum. How much more money does Lisa have left than David?

10 points

100 − 80 =

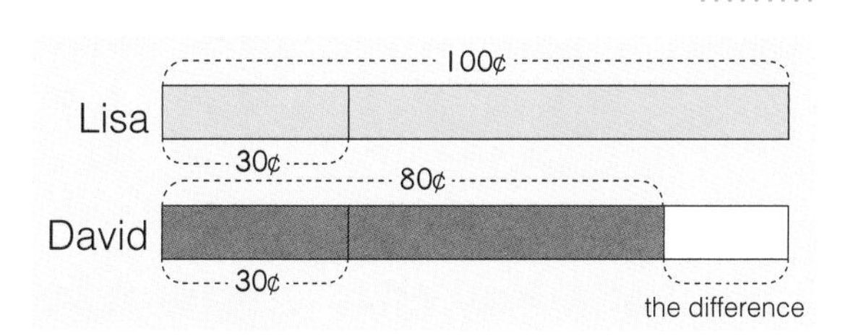

〈Ans.〉

There were 40 red flowers and 30 yellow flowers in our flowerbed. Mother picked 15 flowers of each color. How many more red flowers than yellow flowers are left?

10 points

40 − 30 =

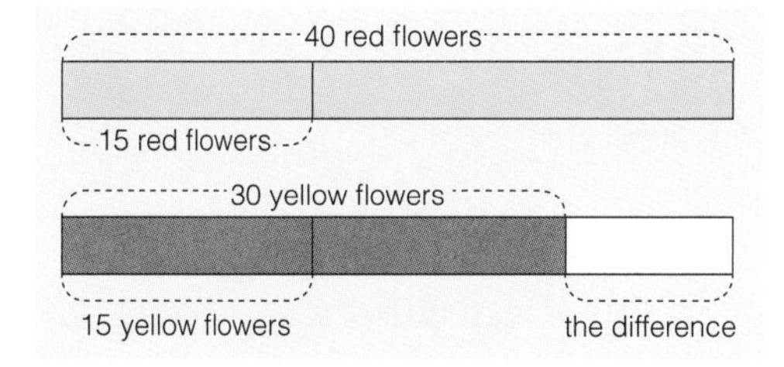

〈Ans.〉

You had 50 red candies and 20 blue candies. You ate 15 candies each and your stomach is sore. How many more red candies than blue candies do you have left?

10 points

〈Ans.〉

Sam had 90¢ and Sally had 70¢. They each bought 30¢ worth of gum. How much more money does Sam have remaining than Sally?

10 points

〈Ans.〉

Once you get the trick of it, it's not so bad, right?

Mixed Calculations

Level ★★★

Date / /

Name

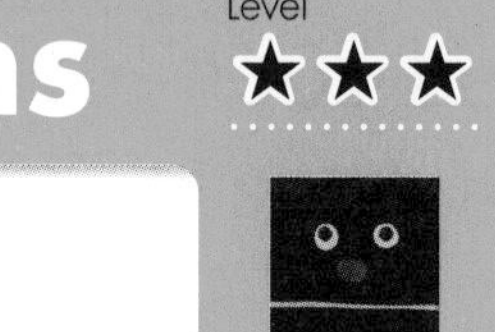

Score /100

1 Some ducks were swimming around the pond. 3 ducks got out to take a nap. Then 2 ducks got out to dry off. Now there are 10 ducks remaining in the pond. How many ducks were in the pond at first?

10 points

$10+2=12$

$12+3=$

10 ducks remaining | 2 ducks | 3 ducks | at first

⟨**Ans.**⟩ ______________

2 Deb got a box of oranges. Yesterday, her family ate 6 and today they ate 7. There are 17 oranges left in the box. How many oranges were in the box at first?

10 points

⟨**Ans.**⟩ ______________

3 During recess, a bunch of children jumped rope. 4 girls joined in, and then 7 boys joined in. Now there are 23 children jumping rope. How many children were jumping rope at first?

10 points

$23-7=16$

$16-4=$

23 children now | at first | 4 girls | 7 boys

⟨**Ans.**⟩ ______________

4 For her birthday, Ella got 13 stickers from your sister and 24 from your brother. Altogether, she now has 100 stickers. How many stickers did she have before?

10 points

⟨**Ans.**⟩ ______________

5 A bus stopped at the mall and let 4 people off. 6 people got on. Now there are 20 people on the bus. How many people were on the bus at first? 15 points

$20 - 6 = 14$

$14 + 4 =$

〈Ans.〉 ____________

6 During the year, 4 students left Kate's class, and 6 students joined. Now there are 36 students in her class. How many students were in her class at the beginning of the year? 15 points

〈Ans.〉 ____________

7 You have sheets of colored paper for crafts class. Today, you got 20 more from your teacher, and used 10. Now, you have 15 sheets of colored paper left. How many sheets did you have before class today? 15 points

$15 + 10 = 25$

$25 - 20 =$

〈Ans.〉 ____________

8 Yesterday, Julia got 50 beads from her mother. Today, she made a bracelet with 40 beads, and she has 130 left. How many beads did she have at the beginning? 15 points

〈Ans.〉 ____________

These problems are getting tough, but I know you can do it.

Mixed Calculations

Level ★★★

Date / /

Name

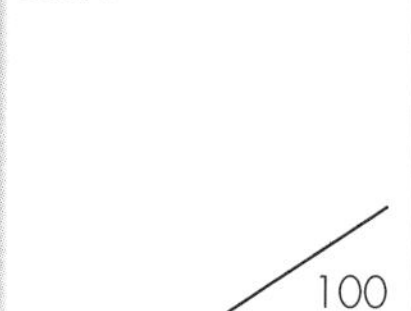
Score /100

1 Karen lined up with the other children for ice cream in the cafeteria. In front of Karen, there are 5 children. Behind her, there are 7 children. How many children are in line altogether?

10 points

$\boxed{5} + \boxed{7} + 1 = \boxed{13}$

〈**Ans.**〉

2 During dodgeball, George lined up against the wall. There were 11 children to his left, and 9 children to his right. How many children were playing dodgeball?

15 points

$11 + 9 + 1 =$

〈**Ans.**〉

3 A long line was waiting for tickets to the movie. In front of Rafael, there were 13 people. There were also 13 people behind him. How many people were in line?

15 points

〈**Ans.**〉

4 A bunch of hats hung on a row of hat pegs. There were 5 hats to the left of David's hat and 8 hats to the right. How many hats were there in all? 15 points

⟨Ans.⟩ ____________

5 At the toll booth, there was a long line of vehicles. In front of the bus there were 12 cars, and behind it were 9 more cars. How many cars were in line altogether? 15 points

⟨Ans.⟩ ____________

6 A line of children was waiting for their flu shots. In front of Sam there were 6 children and behind him were 7 more. How many children were waiting in all? 15 points

⟨Ans.⟩ ____________

7 I have a pile of books. There are 8 books on top of my picture book, and 4 books under it. How many books are in my pile? 15 points

⟨Ans.⟩ ____________

Who knew lines could be so tricky?
Do your best!

29 Mixed Calculations

Level ★★★

Date / /

Name

Score /100

1 The teacher made everyone put their hats on the rack. Lilly's hat is seventh from the left and fifth from the right. How many hats are there?

10 points

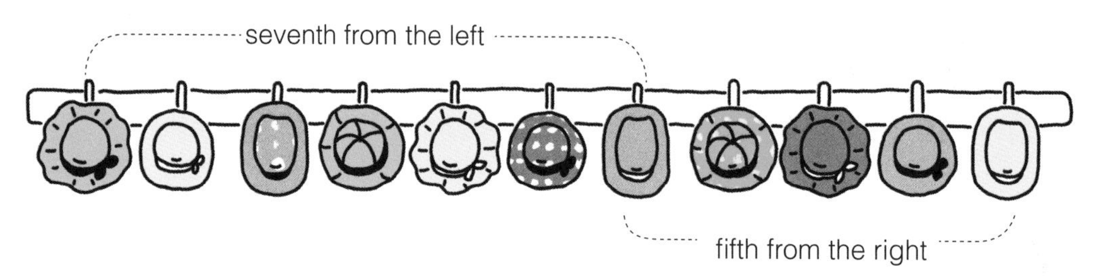

7 + 5 − 1 =

〈**Ans.**〉

2 If Aesop's Fables is the tenth book from the left and the seventeenth book from the right on the class bookshelf, how many books are on the shelf?

15 points

〈**Ans.**〉

3 Adam is eighth in line for the pool. He is twelfth from the back. How many children are in line for the pool today?

15 points

〈**Ans.**〉

4 11 vehicles are waiting to be inspected. Don's truck is sixth from the front. What number vehicle is he from the back? 15 points

$11 - 6 + 1 =$

〈**Ans.**〉 ______________

5 All 25 children in the class put up their drawings in a row. My drawing was tenth from the left. What number drawing from the right was it? 15 points

〈**Ans.**〉 ______________

6 18 people are in line at the bank. Harry's mother is seventh from the front. What number from the back was Harry's mother? 15 points

〈**Ans.**〉 ______________

7 There are 80 stone steps at the church. Andrew is at the twenty-fifth stone step from the bottom. What number step from the top is he? 15 points

〈**Ans.**〉 ______________

You're doing really well. Keep it up!

30 Mixed Calculations

Level ★★★

Date / /

Name

Score /100

1 Everyone is in line to get their grades on their homework. Teddy is seventh from the front and Tom is 2 people in front of Teddy. What number from the front is Tom? 10 points

7 − 2 =

〈**Ans.**〉

2 A new video game is coming out today, and there is a line in front of the store. Jin is thirteenth from the front and Rob is 5 people in front of Jin. What number from the front of the line is Rob? 10 points

〈**Ans.**〉

3 It is raining today, and all of the umbrellas are in a row at the front of the class. Sara's umbrella is twelfth from the left and Kim's is 3 umbrellas to the left of Sara's. What number umbrella is Kim's from the left? 15 points

〈**Ans.**〉

4 The crafts teacher put everyone's craft in a row. Tim's is sixteenth from the left. Anna's is 5 to the left of Tim's. What number craft is Anna's from the left? 15 points

〈**Ans.**〉

5 The teacher put the class in a row for a math exercise. Kate is fifteenth from the right, and Sally is 6 people to the right of Kate. What number from the right is Sally?

10 points

⟨**Ans.**⟩

6 I have a big pile of books on my desk. The book about cats is ninth from the bottom, and the comic book is two books closer to the bottom than the book about cats. What number book from the bottom is the comic book?

10 points

⟨**Ans.**⟩

7 Teacher has a big pile of books on his desk, too. The book about vehicles is seventh from the bottom, and the book about animals is 3 books below the book about vehicles. What number book from the bottom is the book about animals?

10 points

⟨**Ans.**⟩

8 Kim lives on the ninth floor. Diana lives 3 floors above Kim's room. On what floor does Diana live?

10 points

$9 + 3 =$

⟨**Ans.**⟩

9 Sam stopped on the thirteenth stair from the bottom. Michael is 4 stairs above Sam. Michael is on what number stair from the bottom?

10 points

⟨**Ans.**⟩

Now it's time for tables and graphs!

31 Tables & Graphs

Level ★★

Date / /

Name

Score /100

1 You went around the classroom and asked everyone for their favorite fruit. Below is what they answered.

10 points per question

(1) The apple is the favorite fruit of how many people?

( )

(2) The pineapple is the favorite fruit of how many people?

( )

(3) The peach is the favorite fruit of how many people?

( )

(4) The orange is the favorite fruit of how many people?

( )

(5) The banana is the favorite fruit of how many people?

( )

2 Now, let's tabulate the results of the favorite fruit question.

(1) Write the missing numbers in the table below. 5 points per box

Table: Favorite Fruit

Favorite Fruit	Apple	Pineapple	Peach	Orange	Banana
Number of People	5				

(2) Which fruit is the favorite of the most people? 5 points

( )

3 You went around a parking lot and counted all the different types of vehicles.

(1) Count each type of vehicle and write the numbers in the table below. 5 points per box

Table: Number of Vehicles in the Parking Lot

Vehicles	Car	Truck	Bus	Motorbike
Number of Vehicles				

(2) Of which vehicle is there the most? 5 points

( )

1 Use the "Favorite Fruit" table to fill in the graph below.

Table: Favorite Fruit

Favorite Fruit	Apple	Pineapple	Peach	Orange	Banana
Number of People	5	4	3	4	2

(1) Finish filling in the names of the fruit at the bottom of the new graph. 15 points for completion

(2) Draw an × in the column for every person that has each fruit as his or her favorite. 15 points for completion

(3) Which fruit is the favorite of the most people? 10 points

( )

(4) Which fruit is the favorite of the least people? 10 points

( )

Graph: Favorite Fruit

×				
×				
×				
×				
×				
Apple	Pineapple	Peach		

2 Use the "Number of Vehicles in the Parking Lot" table to fill in the graph below.

Table: Number of Vehicles in the Parking Lot

Vehicles	Car	Truck	Bus	Motorbike
Number of Vehicles	6	5	3	7

(1) Finish filling in the names of the vehicles at the bottom of the graph. 10 points for completion

(2) Draw an × in the correct column for every vehicle you counted. 10 points for completion

(3) Of which vehicle is there the most? 10 points

( )

(4) Of which vehicle is there the least? 10 points

( )

(5) How many more motorbikes are there than buses? 10 points

( )

Graph: Number of Vehicles in the Parking Lot

Car	Truck		

A table has numbers in it, and a graph doesn't!

33 Tables & Graphs

Level ★★★

Date / /

Name

Score /100

1 You went around the classroom and asked everyone for their favorite subject. Below is what they answered.

10 points per question

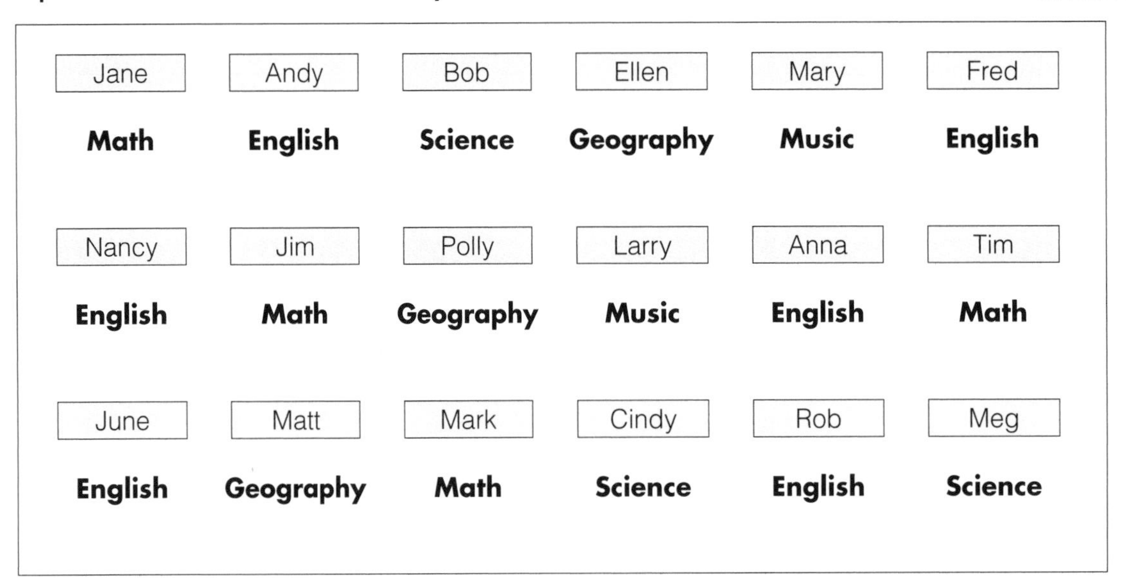

Jane	Andy	Bob	Ellen	Mary	Fred
Math	**English**	**Science**	**Geography**	**Music**	**English**
Nancy	Jim	Polly	Larry	Anna	Tim
English	**Math**	**Geography**	**Music**	**English**	**Math**
June	Matt	Mark	Cindy	Rob	Meg
English	**Geography**	**Math**	**Science**	**English**	**Science**

(1) How many people chose math as their favorite subject?

( )

(2) How many people chose English as their favorite subject?

( )

(3) How many people chose science as their favorite subject?

( )

(4) How many people chose geography as their favorite subject?

( )

(5) How many people chose music as their favorite subject?

( )

2 Use the "Favorite Subject" table to fill in the graph below.

Table: Favorite Subject

Favorite Subject	Math	English	Science	Geography	Music
Number of People	4	6	3	3	2

(1) Finish filling in the names of the subjects at the bottom of the graph. 15 points for completion

(2) Draw an × in the correct column for every person you counted. 15 points for completion

(3) Which subject is the most people's favorite? 10 points

( )

(4) Which subject is the least people's favorite? 10 points

( )

Graph: Favorite Subject

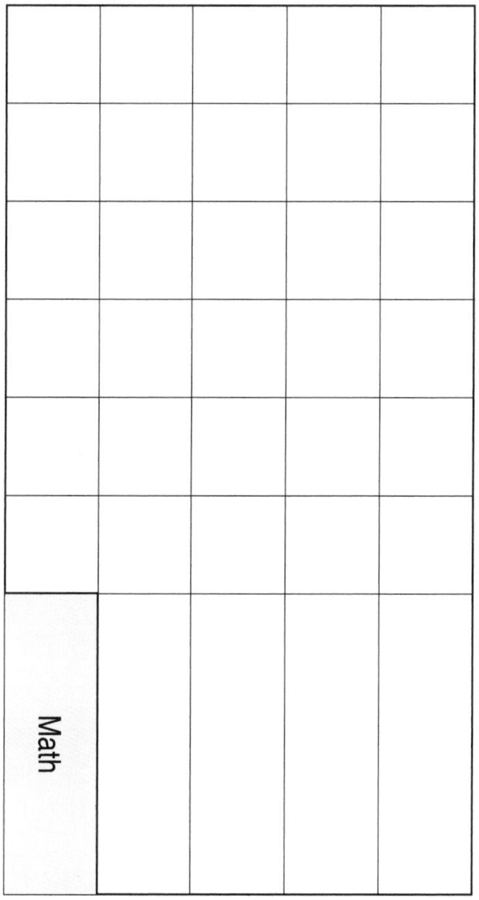

Math				

You did a great job. Now let's review what you've learned.

1 Betty and Juliette were jumping rope. Betty jumped 47 times, and Juliette jumped 66 times. Who jumped more times? How many more times did she jump? 10 points

〈**Ans.**〉

2 There are 101 students in grade 1 of Anna's school. There are 13 fewer students in grade 2. How many students are in grade 2? 10 points

〈**Ans.**〉

3 In your class today, there were 96 students each sitting on a chair. There are still 24 chairs remaining. How many chairs are there in all? 10 points

〈**Ans.**〉

4 There were 16 people on the bus today. At the stop for the hospital, 8 people got off of the bus and 5 got on. How many fewer people are on the bus now? 10 points

〈**Ans.**〉

5 Kim weighs 34 kilograms and Sally weighs 18 kilograms. How much more does Kim weigh? 10 points

〈**Ans.**〉

6 Dave had 70 building blocks. This morning, his father gave him some more. Now, he has 130 blocks. How many blocks did his father give him? 10 points

〈**Ans.**〉

7 At school, there are 140 seats in the auditorium. The auditorium has 50 more seats than the cafeteria. How many seats are in the cafeteria? 10 points

〈**Ans.**〉

8 You used 1 foot 9 inches of tape. If the tape had 4 feet and 11 inches before you started, how much tape is left? 10 points

〈**Ans.**〉

9 June weighed 47 pounds last year. Her weight increased 8 pounds this year. How much does she weigh now? 10 points

〈**Ans.**〉

10 There were 30 oranges on the dish. You ate 7 oranges yesterday and then 9 oranges today. How many oranges are left? 10 points

〈**Ans.**〉

Are you remembering everything? Good!

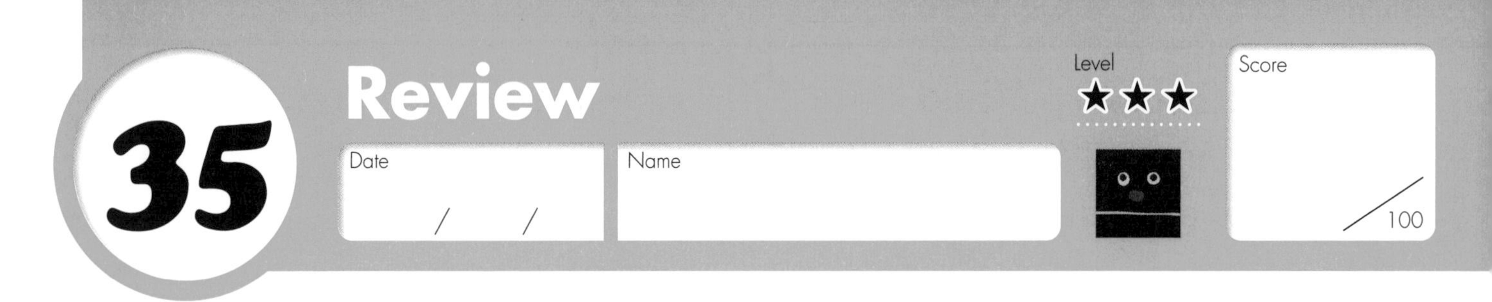

1 In the back of the cafeteria, there is a big box with 42 eggs and a small box with 24 eggs. How many eggs are there in the back of the cafeteria?

10 points

〈Ans.〉

2 The pound has 21 dogs and 16 cats right now. Do they have more dogs or cats? How many more?

10 points

〈Ans.〉

3 Tammy and Nathalie are making ornaments. The red string they are using is 15 meters 5 centimeters long. The blue string is 7 meters long. How much longer is the red string?

10 points

〈Ans.〉

4 You bought some donuts at the store yesterday. Today, you ate 3 of them. You still have 6 donuts remaining. How many donuts did you buy?

10 points

〈Ans.〉

5 Nan picked up 120 acorns, and her brother picked up 75 acorns. How many more acorns did Nan pick up? 10 points

〈**Ans.**〉

6 At Grandfather's farm, there are 76 cows and 105 horses. How many more horses than cows does he have? 10 points

〈**Ans.**〉

7 In our classroom, the red tape is 1 meter 45 centimeters long. The red tape is 29 centimeters longer than the white tape. How long is the white tape? 10 points

〈**Ans.**〉

8 The class is lined up for free ice cream. Ken is seventeenth from the front, and nineteenth from the back. How many children are in line? 15 points

〈**Ans.**〉

9 Ben weighs 39 kilograms and Ann weighs 25 kilograms. How much more does Ben weigh? 15 points

〈**Ans.**〉

Almost there! Nice job!

36 Review

Level ★★★

Date / /　Name　Score /100

1. Mary is reading a storybook about fairies. She read 72 pages and there are 48 pages remaining. How many pages does the book have? 10 points

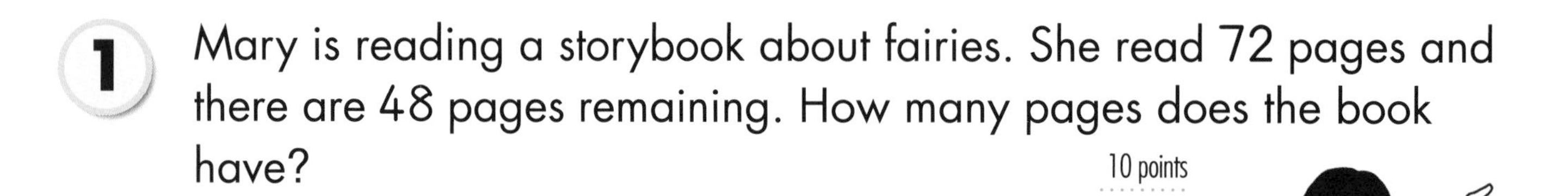

〈Ans.〉

2. In crafts class, Amy got 62 sheets of colored paper and her sister got 48. They each used 7 sheets to make paper cranes. How many more sheets does Amy have now? 10 points

〈Ans.〉

3. The blue tape in father's drawer is 12 feet 7 inches long. The red tape is 6 feet long. How much longer is the blue tape than the red tape? 10 points

〈Ans.〉

4. We have two ropes in our backyard. One is 2 meters 54 centimeters long, and the other is 1 meter 33 centimeters long. If we connect them, how long will the new rope be? 10 points

〈Ans.〉

5. Jim counted 92 dots on the picture of the cheetah. His brother counted 56 dots on his picture of a dalmatian. How many more dots did Jim count than his brother? 10 points

〈Ans.〉

6 The cafeteria has 76 pounds of ham and 68 pounds of bacon. How many more pounds of ham does the cafeteria have? 10 points

〈**Ans.**〉

7 Use the graph here to answer the questions below. 10 points per question

(1) Which fruit is the favorite of the most people?

( )

(2) Which fruit is the favorite of the least people?

( )

(3) How many more people like apples better than oranges?

( )

(4) How many fewer people like grapes better than strawberries?

( )

Graph: Favorite Fruit

Apple	Strawberry	Orange	Grapes	Peach
	×			
×	×			
×	×			×
×	×	×		×
×	×	×	×	×
×	×	×	×	×

Answer Key Grade 2 Word Problems

1 Review
pp 2, 3

1. 4 + 5 = 9 **Ans.** 9 children
2. 7 + 5 = 12 **Ans.** 12 pencils
3. 10 − 3 = 7 **Ans.** 7 apples
4. 11 − 6 = 5 **Ans.** 5 white flowers
5. 8 − 4 = 4
   Ans. He has 4 more apples.
6. 6 + 8 = 14 **Ans.** 14 pencils
7. 9 + 6 = 15 **Ans.** 15 chairs
8. 15 − 8 = 7 **Ans.** 7 cookies
9. 6 + 3 + 5 = 14 **Ans.** 14 people
10. 10 − 5 = 5 **Ans.** Fifth

2 Review
pp 4, 5

1. 5 + 6 = 11 **Ans.** 11 balls
2. 8 − 3 = 5 **Ans.** 5 girls
3. 4 + 2 = 6 **Ans.** 6 pencils
4. 3 + 1 = 4 **Ans.** Fourth
5. 8 − 3 = 5 **Ans.** 5 bottles
6. 9 − 4 = 5 **Ans.** 5 chairs
7. 4 + 5 = 9 **Ans.** 9 cookies
8. 8 − 5 = 3 **Ans.** 3 red stickers
9. 4 + 3 − 2 = 5 **Ans.** 5 dimes
10. 7 − 4 = 3 **Ans.** 3 hats

3 Addition & Subtraction
pp 6, 7

1. 20 + 30 = 50 **Ans.** 50 pages
2. 50 − 30 = 20 **Ans.** 20 pennies
3. 18 + 3 = 21 **Ans.** 21 girls
4. 21 − 3 = 18 **Ans.** 18 girls
5. 36 + 8 = 44 **Ans.** 44 strawberries
6. 25 − 6 = 19 **Ans.** 19 postcards
7. 45 + 8 = 53 **Ans.** 53 pears
8. 43 − 7 = 36 **Ans.** 36 apples
9. 26 + 7 = 33 **Ans.** 33 fish
10. 24 − 9 = 15 **Ans.** 15 cats

4 Addition & Subtraction
pp 8, 9

1. 15 + 17 = 32 **Ans.** 32 cookies
2. 26 − 17 = 9 **Ans.** 9 dimes
3. 24 + 16 = 40 **Ans.** 40 ducks
4. 24 − 18 = 6 **Ans.** 6 cows
5. 38 + 14 = 52 **Ans.** 52 pencils
6. 40 − 35 = 5 **Ans.** 5 notebooks
7. 18 + 27 = 45 **Ans.** 45 stickers
8. 36 − 28 = 8 **Ans.** 8 tangerines
9. 46 + 14 = 60 **Ans.** 60 eggs
10. 32 − 18 = 14 **Ans.** 14 eggs

5 Addition & Subtraction
pp 10, 11

1. 12 − 5 = 7 **Ans.** He has 7 more fish.
2. 12 − 7 = 5
   Ans. He has 5 fewer apples.
3. 25 − 18 = 7
   Ans. The zoo has 7 more foxes.
4. 62 − 45 = 17
   Ans. Maria jumped 17 more times.
5. 80 − 74 = 6
   Ans. We made 6 fewer white airplanes.
6. 120 − 98 = 22
   Ans. There were 22 fewer girls.
7. 145 − 87 = 58
   Ans. I have 58 fewer storybooks.
8. 108 − 96 = 12
   Ans. There were 12 more men.

9 $120 - 95 = 25$

Ans. They had 25 more red roses.

10 $114 - 98 = 16$

Ans. Tom folded 16 fewer cranes.

6 Length

pp 12, 13

1 4 in. + 5 in. = 9 in. **Ans.** 9 in.

2 6 in. + 8 in. = 14 in. **Ans.** 14 in.

3 16 in. + 14 in. = 30 in. **Ans.** 30 in.

4 5 ft. + 7 ft. = 12 ft. **Ans.** 12 ft.

5 8 ft. + 11 ft. = 19 ft. **Ans.** 19 ft.

6 16 in. − 12 in. = 4 in. **Ans.** 4 in.

7 18 in. − 11 in. = 7 in. **Ans.** 7 in.

8 25 in. − 13 in. = 12 in. **Ans.** 12 in.

9 12 ft. − 8 ft. = 4 ft. **Ans.** 4 ft.

10 42 ft. − 14 ft. = 28 ft. **Ans.** 28 ft.

7 Length

pp 14, 15

1 1 ft. 2 in. + 3 in. = 1 ft. 5 in.

Ans. 1 ft. 5 in.

2 2 ft. 5 in. + 5 in. = 2 ft. 10 in.

Ans. 2 ft. 10 in.

3 1 ft. 7 in. + 4 in. = 1 ft. 11 in.

Ans. 1 ft. 11 in.

4 4 ft. 7 in. + 4 ft. = 8 ft. 7 in.

Ans. 8 ft. 7 in.

5 2 ft. 7 in. + 5 ft. = 7 ft. 7 in.

Ans. 7 ft. 7 in.

6 1 ft. 10 in. + 1 ft. 1 in. = 2 ft. 11 in.

Ans. 2 ft. 11 in.

7 2 ft. 8 in. + 1 ft. 2 in. = 3 ft. 10 in.

Ans. 3 ft. 10 in.

8 3 ft. 5 in. + 1 ft. 6 in. = 4 ft. 11 in.

Ans. 4 ft. 11 in.

9 2 ft. 3 in. + 1 ft. 8 in. = 3 ft. 11 in.

Ans. 3 ft. 11 in.

10 5 ft. 7 in. + 4 ft. 1 in. = 9 ft. 8 in.

Ans. 9 ft. 8 in.

8 Length

pp 16, 17

1 1 ft. 11 in. − 10 in. = 1 ft. 1 in.

Ans. 1 ft. 1 in.

2 2 ft. 9 in. − 5 in. = 2 ft. 4 in.

Ans. 2 ft. 4 in.

3 2 ft. 10 in. − 8 in. = 2 ft. 2 in.

Ans. 2 ft. 2 in.

4 8 ft. 9 in. − 5 ft. = 3 ft. 9 in.

Ans. 3 ft. 9 in.

5 42 ft. − 16 ft. = 26 ft. **Ans.** 26 ft.

6 3 ft. 9 in. − 2 ft. 5 in. = 1 ft. 4 in.

Ans. 1 ft. 4 in.

7 4 ft. 10 in. − 2 ft. 7 in. = 2 ft. 3 in.

Ans. 2 ft. 3 in.

8 50 ft. 8 in. − 15 ft. 5 in. = 35 ft. 3 in.

Ans. 35 ft. 3 in.

9 7 ft. 11 in. − 4 ft. 6 in. = 3 ft. 5 in.

Ans. 3 ft. 5 in.

10 5 ft. 9 in. − 2 ft. 3 in. = 3 ft. 6 in.

Ans. 3 ft. 6 in.

9 Length

pp 18, 19

1 7 cm + 8 cm = 15 cm **Ans.** 15 cm

2 (1) 15 cm + 12 cm = 27 cm

Ans. 27 cm

(2) 4 m + 3 m = 7 m **Ans.** 7 m

(3) 6 cm + 5 cm = 11 cm **Ans.** 11 cm

(4) 15 cm + 5 cm = 20 cm

Ans. 20 cm

(5) 1 m + 3 m = 4 m **Ans.** 4 m

(6) 11 cm + 2 cm = 13 cm

Ans. 13 cm

3 15 cm − 7 cm = 8 cm **Ans.** 8 cm

4 (1) 20 cm − 8 cm = 12 cm

Ans. 12 cm

(2) 15 cm − 4 cm = 11 cm

Ans. 11 cm

(3) 16 m − 3 m = 13 m **Ans.** 13 m

(4) 54 cm − 8 cm = 46 cm

Ans. 46 cm

(5) 25 m − 10 m = 15 m **Ans.** 15 m

(6) 70 m − 15 m = 55 m **Ans.** 55 m

10 Length

pp 20, 21

1 1 m 25 cm + 15 cm = 1 m 40 cm

Ans. 1 m 40 cm

2 2 m 47 cm + 22 cm = 2 m 69 cm

Ans. 2 m 69 cm

3 1 m 18 cm + 74 cm = 1 m 92 cm

Ans. 1 m 92 cm

4 4 m 27 cm + 2 m = 6 m 27 cm

Ans. 6 m 27 cm

5 5 m 84 cm + 3 m = 8 m 84 cm

Ans. 8 m 84 cm

6 1 m 46 cm + 2 m 33 cm = 3 m 79 cm

Ans. 3 m 79 cm

7 2 m 13 cm + 1 m 74 cm = 3 m 87 cm

Ans. 3 m 87 cm

8 1 m 38 cm + 3 m 41 cm = 4 m 79 cm

Ans. 4 m 79 cm

9 2 m 17 cm + 1 m 48 cm = 3 m 65 cm

Ans. 3 m 65 cm

10 1 m 73 cm + 1 m 25 cm = 2 m 98 cm

Ans. 2 m 98 cm

11 Length

pp 22, 23

1 1 m 58 cm − 1 m 11 cm = 47 cm

Ans. 47 cm

2 2 m 67 cm − 1 m 23 cm = 1 m 44 cm

Ans. 1 m 44 cm

3 6 m 72 cm − 3 m 43 cm = 3 m 29 cm

Ans. 3 m 29 cm

4 2 m 84 cm − 1 m 31 cm = 1 m 53 cm

Ans. 1 m 53 cm

5 25 m 65 cm − 7 m 35 cm = 18 m 30 cm

Ans. 18 m 30 cm

6 3 m 98 cm − 2 m 19 cm = 1 m 79 cm

Ans. 1 m 79 cm

7 55 m 75 cm − 50 m 50 cm = 5 m 25 cm

Ans. 5 m 25 cm

8 27 m 65 cm − 10 m 55 cm = 17 m 10 cm

Ans. 17 m 10 cm

9 5 m 86 cm − 2 m 52 cm = 3 m 34 cm

Ans. 3 m 34 cm

10 6 m 73 cm − 2 m 54 cm = 4 m 19 cm

Ans. 4 m 19 cm

12 Weight

pp 24, 25

1 2 lb. + 5 lb. = 7 lb. **Ans.** 7 lb.

2 1 lb. + 14 lb. = 15 lb. **Ans.** 15 lb.

3 25 lb. + 12 lb. = 37 lb. **Ans.** 37 lb.

4 74 lb. + 11 lb. = 85 lb. **Ans.** 85 lb.

5 55 lb. + 15 lb. = 70 lb. **Ans.** 70 lb.

6 23 lb. − 9 lb. = 14 lb. **Ans.** 14 lb.

7 87 lb. − 42 lb. = 45 lb. **Ans.** 45 lb.

8 51 lb. − 49 lb. = 2 lb. **Ans.** 2 lb.

9 87 lb. − 35 lb. = 52 lb. **Ans.** 52 lb.

10 53 lb. − 24 lb. = 29 lb. **Ans.** 29 lb.

13 Weight

pp 26, 27

1 3 kg + 2 kg = 5 kg **Ans.** 5 kg

2 5 kg + 20 kg = 25 kg **Ans.** 25 kg

3 27 kg + 15 kg = 42 kg **Ans.** 42 kg

4 52 kg + 24 kg = 76 kg **Ans.** 76 kg

5 $31\text{ kg} + 55\text{ kg} = 86\text{ kg}$ **Ans.** 86 kg
6 $12\text{ kg} - 8\text{ kg} = 4\text{ kg}$ **Ans.** 4 kg
7 $34\text{ kg} - 15\text{ kg} = 19\text{ kg}$ **Ans.** 19 kg
8 $56\text{ kg} - 35\text{ kg} = 21\text{ kg}$ **Ans.** 21 kg
9 $36\text{ kg} - 27\text{ kg} = 9\text{ kg}$ **Ans.** 9 kg
10 $50\text{ kg} - 24\text{ kg} = 26\text{ kg}$ **Ans.** 26 kg

14 Weight

pp 28, 29

1 $5\text{ kg} + 7\text{ kg} = 12\text{ kg}$ **Ans.** 12 kg
2 1 lb. + 10 lb. = 11 lb. **Ans.** 11 lb.
3 $57\text{ kg} + 13\text{ kg} = 70\text{ kg}$ **Ans.** 70 kg
4 $43\text{ kg} + 11\text{ kg} = 54\text{ kg}$ **Ans.** 54 kg
5 76 lb. + 51 lb. = 127 lb. **Ans.** 127 lb.
6 18 lb. − 13 lb. = 5 lb. **Ans.** 5 lb.
7 $44\text{ kg} - 28\text{ kg} = 16\text{ kg}$ **Ans.** 16 kg
8 50 lb. − 27 lb. = 23 lb. **Ans.** 23 lb.
9 $17\text{ kg} - 15\text{ kg} = 2\text{ kg}$ **Ans.** 2 kg
10 75 lb. − 7 lb. = 68 lb. **Ans.** 68 lb.

15 Addition or Subtraction

pp 30, 31

1 $4 + 5 = 9$ **Ans.** 9 children
2 $6 + 3 = 9$ **Ans.** 9 ducks
3 $3 + 5 = 8$ **Ans.** 8 chickens
4 $9 - 4 = 5$ **Ans.** 5 people
5 $8 - 3 = 5$ **Ans.** 5 people
6 (1) $2 + 3 = 5$ **Ans.** 5 pigeons
(2) $16 - 5 = 11$ **Ans.** 11 pigeons
7 (1) $5 + 3 = 8$ **Ans.** 8 apples
(2) $18 - 8 = 10$ **Ans.** 10 apples
8 (1) $4 + 3 = 7$ **Ans.** 7 children
(2) $18 + 7 = 25$ **Ans.** 25 children

16 Addition or Subtraction

pp 32, 33

1 (1) $4 + 5 = 9$ **Ans.** 9 eggs
(2) $20 - (4 + 5) = 11$ **Ans.** 11 eggs
2 $100 - (30 + 60) = 10$ **Ans.** 10¢
3 $40 - (15 + 9) = 16$ **Ans.** 16 stickers
4 $30 - (8 + 6) = 16$ **Ans.** 16 people
5 $19 - (5 + 4) = 10$ **Ans.** 10 sparrows
6 $35 - (10 + 8) = 17$ **Ans.** 17 cards
7 $20 - (6 + 6) = 8$ **Ans.** 8 bars
8 $50 - (9 + 8) = 33$
Ans. 33 strawberries

17 Addition or Subtraction

pp 34, 35

1 $5 + 3 = 8$ **Ans.** 8 sheets
2 $6 + 4 = 10$ **Ans.** 10 flowers
3 $5 + 2 = 7$ **Ans.** 7 apples
4 $8 + 5 = 13$ **Ans.** 13 oranges
5 $20 + 38 = 58$ **Ans.** 58 sheets
6 $75 + 85 = 160$ **Ans.** 160 pages
7 $95 + 35 = 130$ **Ans.** 130 lollipops
8 $75 + 25 = 100$ **Ans.** 100 stickers
9 2 m 30 cm + 60 cm = 2 m 90 cm
Ans. 2 m 90 cm
10 $35 + 85 = 120$ **Ans.** 120 bags

18 Addition or Subtraction

pp 36, 37

1 $8 - 5 = 3$ **Ans.** 3 sheets
2 $10 - 6 = 4$ **Ans.** 4 sparrows
3 $7 - 2 = 5$ **Ans.** 5 candles
4 $9 - 5 = 4$ **Ans.** 4 apples
5 $12 - 10 = 2$ **Ans.** 2 fish
6 $14 - 8 = 6$ **Ans.** 6 postcards
7 $24 - 8 = 16$ **Ans.** 16 pigeons
8 $150 - 80 = 70$ **Ans.** 70 blueberries
9 $30\text{ cm} - 7\text{ cm} = 23\text{ cm}$ **Ans.** 23 cm
10 $100 - 64 = 36$ **Ans.** 36 eggs

19 Addition or Subtraction

pp 38, 39

1 $8 - 3 = 5$ **Ans.** 5 blue sheets
2 $10 - 6 = 4$ **Ans.** 4 people

3) $12 - 4 = 8$ **Ans.** 8 ships
4) $30 - 10 = 20$ **Ans.** 20 dimes
5) $64 - 28 = 36$ **Ans.** 36 cookies
6) $72 - 35 = 37$ **Ans.** 37 pages
7) $52 - 27 = 25$ **Ans.** 25 eggs
8) $125 - 65 = 60$ **Ans.** 60 blocks
9) $60 - 42 = 18$ **Ans.** 18 apples
10) $141 - 87 = 54$ **Ans.** 54 acorns

20 Addition or Subtraction

pp 40, 41

1) $9 - 4 = 5$ **Ans.** 5 coins
2) $10 - 3 = 7$ **Ans.** 7 tadpoles
3) $8 - 3 = 5$ **Ans.** 5 cranes
4) $100 - 40 = 60$ **Ans.** 60¢
5) $23 - 5 = 18$ **Ans.** 18 goldfish
6) $25 - 9 = 16$ **Ans.** 16 pigeons
7) $38 - 16 = 22$ **Ans.** 22 children
8) $65 - 37 = 28$ **Ans.** 28 pages
9) $72 - 25 = 47$ **Ans.** 47 stickers
10) $130 - 70 = 60$ **Ans.** $60

21 Addition or Subtraction

pp 42, 43

1) $9 + 7 = 16$ **Ans.** 16 apples
2) $20 + 12 = 32$ **Ans.** 32 sheets
3) $16 - 7 = 9$ **Ans.** 9 ducks
4) $45 - 6 = 39$ **Ans.** 39 clips
5) $16\text{ m} - 4\text{ m} = 12\text{ m}$ **Ans.** 12 m
6) $17 - 9 = 8$ **Ans.** 8 yellow flowers
7) $30 - 16 = 14$ **Ans.** 14 cards
8) $25 - 6 = 19$ **Ans.** 19 chicks
9) $82 - 35 = 47$ **Ans.** 47 stickers
10) $120 - 45 = 75$ **Ans.** 75 chocolate chips

22 Addition or Subtraction

pp 44, 45

1) $90 - 50 = 40$ **Ans.** 40¢
2) $90 - 40 = 50$ **Ans.** 50¢
3) $80 - 50 = 30$ **Ans.** 30¢
4) $70 - 30 = 40$ **Ans.** 40¢
5) $80 - 25 = 55$ **Ans.** 55¢
6) $62 - 15 = 47$ **Ans.** 47 pennies
7) $95\text{ cm} - 28\text{ cm} = 67\text{ cm}$ **Ans.** 67 cm
8) $140 - 60 = 80$ **Ans.** 80 beads
9) $124 - 96 = 28$ **Ans.** 28 sheets

23 Addition or Subtraction

pp 46, 47

1) $40 + 30 = 70$ **Ans.** 70¢
2) $70 + 20 = 90$ **Ans.** 90 coins
3) $30 + 45 = 75$ **Ans.** 75¢
4) $35 + 55 = 90$ **Ans.** 90¢
5) $48 + 26 = 74$ **Ans.** 74 red sheets
6) $35 + 17 = 52$ **Ans.** 52 stickers
7) $60 + 90 = 150$ **Ans.** 150 markers
8) $85 + 65 = 150$ **Ans.** 150 pieces
9) $27\text{ cm} + 58\text{ cm} = 85\text{ cm}$ **Ans.** 85 cm

24 Addition or Subtraction

pp 48, 49

1) $130 - 55 = 75$ **Ans.** 75 lilies
2) $61\text{ cm} - 26\text{ cm} = 35\text{ cm}$ **Ans.** 35 cm
3) $85 + 60 = 145$ **Ans.** 145 jump ropes
4) $37 + 14 = 51$ **Ans.** 51 pages
5) $36 + 32 = 68$ **Ans.** 68 years old
6) $102 - 16 = 86$ **Ans.** 86 adults
7) $67 + 28 = 95$ **Ans.** 95 sheets
8) $80 + 25 = 105$ **Ans.** 105¢
9) $69 - 14 = 55$ **Ans.** 55 melons
10) $75 + 8 = 83$ **Ans.** 83 acorns

25 Mixed Calculations

pp 50, 51

1) $5 + 6 = 11$ **Ans.** 11 eggs
2) $6 + 12 = 18$ **Ans.** 18 pencils
3) $25\text{ cm} + 18\text{ cm} = 43\text{ cm}$ **Ans.** 43 cm
4) $18 + 15 = 33$ **Ans.** 33 notes

5 $15+17=32$ **Ans.** 32 oranges

6 $70+50=120$ **Ans.** 120 stickers

7 $75+65=140$ **Ans.** 140 bars

8 $85+43=128$ **Ans.** 128 packs

9 $80+32=112$ **Ans.** 112 oranges

10 75 cm $+ 60$ cm $= 135$ cm
Ans. 135 cm

26 Mixed Calculations
pp 52, 53

1 (1) ⟨Mary⟩ $30-5=25$ **Ans.** 25 sheets
⟨Ellen⟩ $20-5=15$ **Ans.** 15 sheets
(2) $25-15=10$ **Ans.** 10 sheets
(3) $30-20=10$ **Ans.** 10 sheets

2 (1) ⟨Ava⟩ $80-40=40$ **Ans.** 40¢
⟨Ted⟩ $50-40=10$ **Ans.** 10¢
(2) $40-10=30$ **Ans.** 30¢
(3) $80-50=30$ **Ans.** 30¢

3 $100-80=20$ **Ans.** 20¢

4 $40-30=10$ **Ans.** 10 red flowers

5 $50-20=30$ **Ans.** 30 red candies

6 $90-70=20$ **Ans.** 20¢

27 Mixed Calculations
pp 54, 55

1 $10+2=12$, $12+3=15$
Ans. 15 ducks

2 $17+7=24$, $24+6=30$
Ans. 30 oranges

3 $23-7=16$, $16-4=12$
Ans. 12 children

4 $100-24=76$, $76-13=63$
Ans. 63 stickers

5 $20-6=14$, $14+4=18$
Ans. 18 people

6 $36-6=30$, $30+4=34$
Ans. 34 students

7 $15+10=25$, $25-20=5$
Ans. 5 sheets

8 $130+40=170$, $170-50=120$
Ans. 120 beads

28 Mixed Calculations
pp 56, 57

1 $5+7+1=13$ **Ans.** 13 children

2 $11+9+1=21$ **Ans.** 21 children

3 $13+13+1=27$ **Ans.** 27 people

4 $5+8+1=14$ **Ans.** 14 hats

5 $12+9+1=22$ **Ans.** 22 cars

6 $6+7+1=14$ **Ans.** 14 children

7 $8+4+1=13$ **Ans.** 13 books

29 Mixed Calculations
pp 58, 59

1 $7+5-1=11$ **Ans.** 11 hats

2 $10+17-1=26$ **Ans.** 26 books

3 $8+12-1=19$ **Ans.** 19 children

4 $11-6+1=6$ **Ans.** Sixth

5 $25-10+1=16$ **Ans.** Sixteenth

6 $18-7+1=12$ **Ans.** Twelfth

7 $80-25+1=56$ **Ans.** Fifty–sixth

30 Mixed Calculations
pp 60, 61

1 $7-2=5$ **Ans.** Fifth

2 $13-5=8$ **Ans.** Eighth

3 $12-3=9$ **Ans.** Ninth

4 $16-5=11$ **Ans.** Eleventh

5 $15-6=9$ **Ans.** Ninth

6 $9-2=7$ **Ans.** Seventh

7 $7-3=4$ **Ans.** Fourth

8 $9+3=12$ **Ans.** Twelfth

9 $13+4=17$ **Ans.** Seventeenth

31 Tables & Graphs
pp 62, 63

1 (1) 5 people (2) 4 people
(3) 3 people (4) 4 people
(5) 2 people

2 (1)

Favorite Fruit	Apple	Pineapple	Peach	Orange	Banana
Number of People	5	4	3	4	2

(2) Apple

3 (1)

Vehicles	Car	Truck	Bus	Motorbike
Number of Vehicles	6	5	3	7

(2) Motorbike

32 Tables & Graphs

pp 64,65

1 (1) (2)

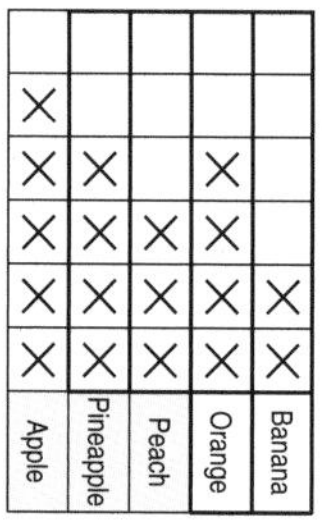

×				
×	×		×	
×	×	×	×	
×	×	×	×	×
×	×	×	×	×
Apple	Pineapple	Peach	Orange	Banana

(3) Apple
(4) Banana

2 (1) (2)

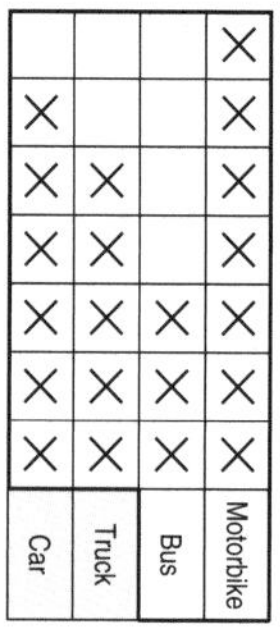

			×
×			×
×	×		×
×	×		×
×	×	×	×
×	×	×	×
×	×	×	×
Car	Truck	Bus	Motorbike

(3) Motorbike
(4) Bus
(5) 4 motorbikes

33 Tables & Graphs

pp 66,67

1 (1) 4 people (2) 6 people
(3) 3 people (4) 3 people
(5) 2 people

2 (1) (2)

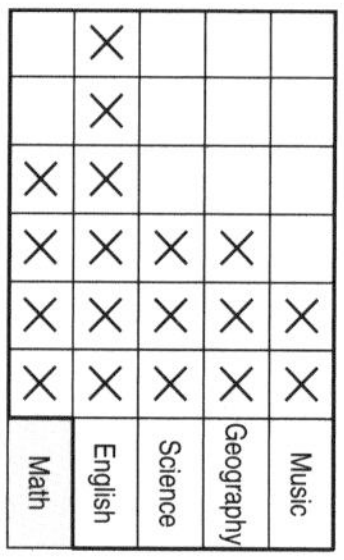

	×			
	×			
×	×			
×	×	×	×	
×	×	×	×	×
×	×	×	×	×
Math	English	Science	Geography	Music

(3) English
(4) Music

34 Review

pp 68,69

1 $66 - 47 = 19$
Ans. Juliette jumped 19 more times.

2 $101 - 13 = 88$ **Ans.** 88 students

3 $96 + 24 = 120$ **Ans.** 120 chairs

4 $8 - 5 = 3$ **Ans.** 3 people

5 34 kg − 18 kg = 16 kg **Ans.** 16 kg

6 $130 - 70 = 60$ **Ans.** 60 blocks

7 $140 - 50 = 90$ **Ans.** 90 seats

8 4 ft. 11 in. − 1 ft. 9 in. = 3 ft. 2 in.
Ans. 3 ft. 2 in.

9 47 lb. + 8 lb. = 55 lb. **Ans.** 55 lb.

10 $30 - (7 + 9) = 14$ **Ans.** 14 oranges

35 Review

pp 70,71

1 $42 + 24 = 66$ **Ans.** 66 eggs

2 $21 - 16 = 5$
Ans. There are 5 more dogs.

3 15 m 5 cm − 7 m = 8 m 5 cm
Ans. 8 m 5 cm

4 $6 + 3 = 9$ **Ans.** 9 donuts

5 $120 - 75 = 45$ **Ans.** 45 acorns

6 $105 - 76 = 29$ **Ans.** 29 horses

7 1 m 45 cm − 29 cm = 1 m 16 cm
Ans. 1 m 16 cm

8 $17 + 19 - 1 = 35$ **Ans.** 35 children

9 39 kg − 25 kg = 14 kg **Ans.** 14 kg

36 Review

pp 72,73

1 $72 + 48 = 120$ **Ans.** 120 pages

2 $62 - 48 = 14$ **Ans.** 14 sheets

3 12 ft. 7 in. − 6 ft. = 6 ft. 7 in.
Ans. 6 ft. 7 in.

4 2 m 54 cm + 1 m 33 cm = 3 m 87 cm
Ans. 3 m 87 cm

5 $92 - 56 = 36$ **Ans.** 36 dots

6 76 lb. − 68 lb. = 8 lb. **Ans.** 8 lb.

7 (1) Strawberry (2) Grapes
(3) 2 people (4) 4 people

